Building Acoustics

Building Acoustics

Marc Asselineau

Peutz & Associates

CRC Press
Taylor & Francis Group
Boca Raton London New York

CRC Press is an imprint of the
Taylor & Francis Group, an **informa** business

CRC Press
Taylor & Francis Group
6000 Broken Sound Parkway NW, Suite 300
Boca Raton, FL 33487-2742

© 2015 by Taylor & Francis Group, LLC
CRC Press is an imprint of Taylor & Francis Group, an Informa business

No claim to original U.S. Government works

Printed on acid-free paper
Version Date: 20150209

International Standard Book Number-13: 978-1-4665-8244-6 (Paperback)

Visit the Taylor & Francis Web site at
http://www.taylorandfrancis.com

and the CRC Press Web site at
http://www.crcpress.com

Contents

Foreword

When I met Marc Asselineau for the first time during the International Congress of Acoustics in Beijing in 1992, we quickly befriended each other because we were happy to find a bit of France at this event that was not heavily attended by Europeans at that time. These were the early years of a young PhD possessed with a passion for music who was destined to enjoy a long career in a large international engineering consulting company. Those were the times when computer models were appearing that could compute the acoustic performances of buildings and the acoustic quality of large spaces. Famous names in architectural acoustics, such as Lothar Cremer, Leo Beranek, and Robert Josse in France, had defined the basis of this discipline in the middle of the 20th century, but one had to wait until the 1980s to see the considerable development of architectural acoustics driven by a strong demand for comfort and quality in built spaces, and then made possible by the abundant choice of products, components, and processes, especially developed for building acoustics. Motivated by the requirements of the European single market and the development of ISO and CEN standards, laboratory test methods have been improved.

However, neither computer models nor standards or test codes nor test reports, as excellent as they may be, provide the one element that is indispensable to the acoustician, the architect, and the contractor to solve their problems, namely, practical experience. It is this particular element that this book presents, and that to my knowledge has no equal apart, partly, from a few courses given by experienced acousticians to a privileged few.

My experience as a consultant at the French Information and Documentation Centre on Noise (CIDB), who meets people experiencing difficulties with noise, shows that the first and perhaps the most difficult part is to try and properly analyze the problem of the client, who comes in the hope of being delivered a prescription (in the medical sense) for a "product" that takes not too much space and is preferably not expensive that will solve everything even before the actual cause can be identified. The distinctive quality of this book is that it tries to prompt the reader to analyze the various situations globally and shows through numerous real life examples the route to understanding and solving the problem. An apprenticeship in acoustics through examples is certainly easier than one gained by learning theory, and it is also more convincing to a number of professionals who work hands-on, such as architects and contractors. This does not prevent the curious minded or the specialist reader from digging deeper into the very complete literature listed at the end of each chapter.

The book can be read in two different ways:

1. The architect or engineer who looks to solve a specific problem can quickly find in the table of contents the chapter of interest and go to the required information without having to read all the preceding text.

2. The beginner in building acoustics can use this as a bedside book: it is easy to read, the style is pleasant and often funny, and the numerous examples of what should be done, and especially what should not be done, show that beside being a scientific discipline acoustics require a good deal of observation and a capacity for situation analysis.

A very novel chapter covers interactions between acoustics and other disciplines such as structural engineering and HVAC engineering that require compromises that are satisfactory to every one of them.

The acoustic quality of rooms is certainly the domain of acoustics that is the most difficult to resolve, as it inevitably reflects the knowledge of the designer, but also his or her response to schools of thought, aesthetical trends, and the opinions of musical critics. The acoustics of a large and prestigious hall represent somewhat personal work that will be either foremostly liked or disliked by the audience. In this respect, a merit of the book is that it illustrates various points of view through real projects, some of which have been undertaken by the author.

I have no doubt that this original book offers a precious service to anyone looking to enter the fascinating world of acoustics and to progress quickly without waiting for their own experience to accumulate.

Jacques Roland

Former Head of the Acoustic and Lighting Department of the French Scientific and Technical Building Centre (CSTB), former president of the European Standardization Technical Committee TC126 "Building Acoustics"

Preface

Many years ago an architect and friend of mine took the opportunity of the inauguration of his latest building to tease his architectural team and his engineers. In the course of this exercise he pointed out that the acoustician really could be a nuisance—he would complain there is too much noise in one place but not enough sound level at another; he would dream of uncoupled structures; and he might even dream of strange, ugly looking surface coverings in rooms. He eventually concluded that the acoustician probably is blind. Not to be outdone, the acoustician retorted that the architect has probably been deaf. Other specialists got their share as well from the structural engineer accused of constantly lightening the superstructure to the HVAC engineer suspected of reverting back to a medieval castle-like façade full of narrow slits. Once the laughter had settled we slowly realized that although we had been working together for quite a number of years, we just did not know much about the other's trade.

Back at the office, reflecting on the exchange I then remembered some questions by respectable university people knowledgeable in the field of physical acoustics, who clearly did not know much about noise reduction, let alone building with an acoustic intent. I also vividly remembered that whenever I was tasked with teaching the basics of noise control in industry, there were bound to be at least a couple of questions from the audience about concert hall acoustics and another about noise control at home!

This particular exchange eventually came to haunt me when I was invited to try and write a book. One of the first tasks was to try and see what others had previously written. There are quite a number of academics who have managed to write large theoretical books, so there was no need to try and fit too many formulas that practically nobody would bother to understand anyway (not only the mathematics but also their domain of validity). Although I have tried to give the basics that a beginner would expect to find, my thoughts have been directed toward colleagues and friends. I have also tried to give an overview of the acoustical challenges that a project may bring forward for several types of construction. I have also included a long list of incidents observed over more than a quarter of century of professional experience to illustrate the point and leave an impression on the reader.

Acknowledgments

This book would not have been possible without the encouragement of quite a few people. To start with I thank Mr. Tony Moore, Senior Editor, Taylor & Francis, for inviting me to write this book.

I acknowledge the support of those of my colleagues from Peutz who encouraged me, especially Mrs. A. Gaulupeau, who provided both technical advice and moral support! I also thank Peutz & Associates through MM J. Granneman (General Director of Peutz) and S. Mercier (Director of Peutz France) who authorized me to write this book during my free time and eventually publish it. I also thank Professor J.G. Migneron (Université Laval, Québec, Canada), who inducted me in the field of building and environmental acoustics, and Senior Engineer P. Heringa (Peutz, Nijmegen, Netherlands) who encouraged me throughout my career.

A special thank you to Mr. Q. Gaulupeau for his skill at the craft of enhancing my drawings.

Special thanks are due to those of my colleagues and directors, especially MM. R. Metkemeijer and K. Ogendoorn, who encouraged over the last 30 years with my involvement in acoustic congresses as they proved to be a permanent source of knowledge.

I would also like to thank several clients and partners as well as friends who pushed me to write; Dr. Chevret (INRS), Dr. Viollon (EDF), and Mrs. Seve (formerly with UGC cinemas).

Finally, I express my thanks to my family who encouraged me in this work.

Chapter 1

Introduction

In the antique world acoustics was considered to be the science of sciences, the one that could explain universal harmony.

Nowadays, it often suffers from an ostentatious separation between what is considered to be the academic world and what is considered to be real life. There are quite a few fields under the name *acoustics*, ranging from infrasound to ultrasound through the audible world, and dealing with different things from practical noise control in a building to sonar on ships to physiological damage to the hearing system.

This book is devoted to the field of building acoustics. Equations have been kept to a minimum, while literature references have been offered for the reader interested to push further his theoretical knowledge.

Experience shows that a good building usually is a project where everybody concerned has been able to make his point and the various issues have been discussed and understood by all interested parties. Acoustics often is part of a global problem that can only be solved by a complete design team fed the relevant data by the end user and accustomed to collaborative work.

Experience has, sometimes painfully, taught us that the first condition for a building project to succeed is for everybody on the design team and the end user team too to try and understand each other's needs. But in order to be able to communicate with the other members of the team, the acoustician prefers not to be obliged to explain again and again some rather basic (to him) principles. Similarly, the other team members love it when they are not reduced to explaining to a dumb acoustician why they so desperately need a lighter structure or a specific thermal insulation. The basic idea object of the exercise, as we currently see it, is not to try and produce one more building acoustics book whose reading is limited to specialists, but to try to provide a nonspecialist reader with keys for understanding each other's requirements.

More to the point, one can be appalled on a regular basis when discovering how rather simple mistakes have managed to reach construction stage or, even worse, commissioning stage (e.g., What about polystyrene as an absorptive material? Do not laugh, it has happened!). Hence, this book is readable by nonspecialists, if only to give them a rough idea of what the problems might be and what should be asked by all involved parties to clarify the matter (after all, this is the spirit of today's sustainable development projects, isn't it?). Examples are given to illustrate the various points of interest. Readers especially interested in the subject (e.g., a junior acoustical engineer, or an architect or engineer in another speciality) can then go on to a more specialized chapter and complete their learning through the relevant literature references.

While performance halls have, of course, found their place in this book, smaller but essential spaces (those whose bad acoustics can ruin your day) are discussed: open-plan offices and restaurants (yes, there are many similarities between them), production facilities, sport facilities, meeting rooms, and transport stations, just to name a few.

While the reader may not emerge as a fully fledged engineer after reading this book, he will come out knowing enough to express his needs to a specialized consultant and to avoid the worst acoustic mistakes should he be left to his own devices.

Chapter 2

Acoustics

2.1 DEFINITION OF ACOUSTICS

Acoustics is defined as the science of sound. The word comes from the Greek *acoustikos*, meaning "for hearing."

Hearing is not limited to the human species. It is crucial in the animal world, as it enables them to detect an incoming danger or help pinpoint prey. More to the point, speech is used by many species, especially humans, for communication.

In the antique world acoustics was sometimes considered the science of sciences. Nowadays, it features numerous widely branches ranging, for example, physical acoustics, building acoustics, underwater acoustics, etc. A quite comprehensive list can be found in [1, 2]. Basically, when dealing with acoustics, three entities are involved: a source, a propagation medium, and a receiver. The source converts energy into vibrations of the propagation medium; it can be, for example, a vibrating rod or a surface. The propagation medium can be solid (e.g., metal, wood), liquid (e.g., water), or gas (e.g., air).

2.2 BRIEF HISTORY

The reader interested in the history of acoustics can find more developed information in references [3–6]. Here is only a brief reminder of a few points of interest.

In the antique world acoustics was sometimes considered the science of sciences. Musical acoustics were studied by scientists and philosophers who dreamed to discover the secrets of world order. During his investigations of musical intervals, Pythagoras (6th century BC) discovered harmonics. Vitruvius wrote a treatise featuring considerations on theatres, including echoes and reverberation (20 BC); he also compared sound propagation to the circles in water. Aristotle explained wave motion as contractions and expansions of air bumping into the one next to it.

The knowledge of the Greeks was passed on to the Romans and the Arabs, with the latter developing during the Middle Ages. Later the Renaissance saw the development of architectural acoustics and musical acoustics again.

Kircher (1602–1680) worked on sound propagation. While he wrongly concluded that no propagation medium was necessary, he also worked in architectural acoustics regarding the focusing by vaulted ceilings and the amplification effect by the bell of brass instruments.

Mersenne (1588–1648) managed to measure the sound propagation speed in 1640 using an artillery gun; he found that attenuation was a function of the distance to the power 2. He also found that the resonance frequency of a string was inversely proportional to its length and proportional to the square root of its tension [7]. Using both a handheld gun (generating high-pitched sounds) and an artillery gun (generating low-pitched sounds)

Gassendi (1592–1655) experimentally proved that speed propagation was independent of frequency. Huygens (1629–1695) showed that sound is an undulatory phenomenon, and Newton developed a mathematical formulation of the propagation of sound with an expression of the sound speed. Sauveur (1653–1716) looked at the composition of sounds and distinguished between the fundamental and its harmonics, whose decomposition defines tone color. He also defined nodes (locations where there is no elongation) and antinodes (locations where the elongation is maximal), and used the term *acoustics* to cover the science of sound. Bernoulli (1700–1782) explained the coexistence of small oscillations too.

With the development of mathematics in the 18th century, serious work was undertaken regarding sound propagation. D'Alembert (1717–1783), Euler (1707–1783), and Lagrange (1736–1813) developed the formal wave propagation equations. And Fourier (1768–1831) proposed his harmonic analysis of sound. Later developments included sound propagation in liquids (Sturm and Colladon in 1827) and solids (Hassenfratz in 1794 and Biot in 1808), as well as membrane or plate vibrations that were experimentally demonstrated by Chladni and theoretically investigated by Germain (1815), Poisson, and Clebsch.

The 19th century saw a significant interest in the perception of sound. Ohm showed that the hearing mechanism could distinguish frequencies (1843), and Corti proposed a model of the inner ear in 1851. Fechner published his *Elements of Psychophysics* (1860), where the relationship between sensation and excitation was investigated. Helmholtz (1821–1894), who was both a physicist and a physiologist, published his physiological theory of music (1877). Physical acoustics were not forgotten as Lord Rayleigh (1842–1919) published his *Theory of Sound* (1877) and Kundt (1839–1894) worked on resonances and stationary waves in tubes.

The 20th century saw an increase of interest in computational matters. As soon as 1901, the Boston Symphony Hall was the first such facility to be designed (by none other than Sabine) using acoustic computations. With the availability of computational methods and new technology, a first attempt at active noise cancellation was made in 1934 by Olsen, and Cremer developed the theory of sound transmission through a wall (1942).

2.3 NOTIONS OF LEVELS

2.3.1 Characterizing Sound

Sound, as perceived by our ears, is made of periodic vibrations of air. It can last for a given duration of time. It can feature a specific pitch. It can be more or less intense. This means that to describe a sound, one must use three dimensions: time (in s), frequency (in Hz), and level (in dB).

The representation of level in dB versus frequency in Hz is known as the *spectrum*.

2.3.2 Sound Level

According to Weber's law, perception varies like the logarithm of excitation. This has led to a logarithmic expression of the sound level L_p, which is expressed as

$$L_p = 10 \log (p^2/p_0^2)$$

where p is the variation of acoustic pressure, and p_0 is the reference corresponding to the smallest perceptible acoustic pressure variation, with

$$p_0 = 2 \times 10^{-5} \text{ Pa}$$

2.3.3 Weightings

The human ear does not perceive all sounds in the same way, depending on their frequency and loudness. This has led to a frequency weighting system standardized over the years by the International Electrotechnical Commission (IEC) [8], based on the Fletcher and Munson equal loudness curves [10]. The A weighting was initially introduced for low-level sounds (up to 40 phons), with higher-level sounds being treated to other weighting curves designated as B, C, and D, the latter being especially devoted to aircraft sound level measurements. Nowadays, both the B and the D curves have disappeared [9], but the C curve, which better takes into account the low-frequency sound levels, is currently used in a few occupational noise regulations. More to the point, it has also found its way into some community noise regulations, for example, in Scandinavian countries [11], where it is used to help define limits for background noise.

Here are a few examples of A-weighted sound level values:

18 dB(A): Woodland area without wind (and without birds singing either!)
30 dB(A): Cinema projection room (empty)
45 dB(A): Workstation with the desktop computer's fan running
70 dB(A): Busy street
90 dB(A): Airport façade with a plane maneuvering at the pier
140 dB(A): 5 m from a jet engine

Now one has probably noted that due to the weighting curve shape, it is possible to achieve a given weighted value with rather different spectrum shapes. For example, a 100 dB(A) value can be achieved with a 100 dB pure tone at 1000 Hz, but also with 126 dB at 63 Hz. In order to avoid the presence of too sharp a tone or a frequency band in a spectrum, one usually specifies noise limits using simultaneously a global A-weighted sound level value and a frequency contour featuring higher levels in the lower-frequency range than in the higher-frequency range. One can either use the noise rating (NR) contour as defined by standard ISO 1996:1971 [12] or the noise criteria (NC) contour as defined by ANSI S12-2-2008 [13].

2.3.4 Addition

Addition of sound level contributions can be performed as a logarithmic addition. Let's add two contributions, L_{p1} and L_{p2}; the resulting sound pressure level L_{ptot} will be

$$L_{ptot} = 10 \log ((10^{\wedge}(L_{p1}/10)) + (10^{\wedge}(L_{p2}/10)))$$

Should one find logarithms unmanageable, all is not lost: it is possible to use a simple table of additions (Table 2.1). Starting with the highest of the two sound level values to be added (which is noted X in Table 2.1), one looks up the difference to the value to be added. A couple of examples are given in Section 2.3.6.2.

By the way, one can derive an important consequence for noise control purposes: Adding a noise source whose contribution is no greater than the original sound level minus 15 dB will not affect the overall sound level value.

2.3.5 Equivalent Sound Levels and Statistical Sound Levels

How does one describe a temporally fluctuating noise? A simple way is to make reference to its energetic value and use the equivalent sound level, given the symbol L_{eq}, which represents a nonfluctuating signal containing as much acoustic energy as the signal under study over the period of time considered. Incidentally, the A-weighted value, which is widely used in surveys, is given the symbol L_{Aeq}.

Table 2.1 Result of adding to a sound level L_{p1} of X dB a second level L_{p2}

Adding X dB and:	Result
X dB	X + 3 dB
X + 1 dB	X + 3.5 dB
X + 2 dB	X + 4.1 dB
X + 3 dB	X + 4.8 dB
X + 4 dB	X + 5.5 dB
X + 5 dB	X + 6.2 dB
X + 6 dB	X + 7.0 dB
X + 7 dB	X + 7.8 dB
X + 8 dB	X + 8.6 dB
X + 9 dB	X + 9.5 dB
X + 10 dB	X + 10.4 dB
X + 11 dB	X + 11.3 dB
X + 12 dB	X + 12.3 dB
X + 13 dB	X + 13.2 dB
X + 14 dB	X + 14.2 dB
X + 15 dB	X + 15.1 dB
X + 16 dB	X + 16.1 dB

Are we done then? Of course not. For the same L_{eq} value the temporal fluctuations of a noise can be very different, and people can be sensitive to such variations. For example, a 65 dB(A) L_{Aeq} value can be reached close to a major highway where it is a continuous rumble, but it can also be reached close to a country road if a single motorcycle passes by [14]! Clearly enough, something else is needed. One then uses the notion of statistical sound levels, given the symbol L_x, which states the sound level reached or exceeded for $x\%$ of the analysis time. Several standards and regulations currently use L_{50} (which will be less than the L_{eq} value for a fluctuating noise), as well as L_{10} (which gives a fair idea of the highest noise levels) and L_{90} (which gives a fair idea of the lowest noise levels), with the difference giving the dynamic of the noise.

What about the fluctuations of noise over a whole day? Legal writers and standard writers have eventually come up with the notion of day-evening-night, where the energetic sum is weighted according to the period of the day (i.e., there is a penalty of 5 dB for evening noise and 10 dB for night noise):

$$L_{den} = 10 \log \{12 \ (10^\wedge(L_d/10)) + 4 \ (10^\wedge((L_e + 5)/10)) + 8 \ (10^\wedge((L_n + 10)/10))$$

where L_d, L_e, and L_n are the day (7:00 a.m to 7:00 p.m.), evening (7:00 p.m. to 11:00 p.m.), and night (11:00 p.m. to 7:00 a.m.) values, respectively, of the noise levels. Please note that while the durations of each period are clearly defined in the European Union, each member state can adjust the corresponding time limits.

2.3.6 Examples

2.3.6.1 Highest Sound Level in Air under Normal Circumstances

What is the highest sound level value that one might theoretically encounter in air? Let's first go back to the definition of a sound level. It is clear that to find the maximum sound pressure level value, one has to find out what the highest variation of sound pressure might be. This variation occurs around the mean atmospheric pressure (1.01325×10^5 Pa) and cannot be negative, so we have $\Delta p_{max} = 1.01325 \times 10^5$ Pa, which corresponds to $L_{pmax} = 194$ dB.

2.3.6.2 Addition

Let's first consider two pieces of equipment, with each of them singly generating 50 dB(A) at a workstation. When both are operating simultaneously, the resulting sound level value will be 53 dB(A).

Let's now consider two pieces of equipment with the noisier one generating 70 dB(A) at a workstation and the other one 50 dB(A). When both are operating simultaneously, the resulting sound level value is 70 dB(A).

Note: This has a consequence when attempting noise reduction. If you do not work on the highest contributing noise source, you will have trouble making any progress.

Let's complicate things a bit further by considering 400 pieces of equipment, with each of them singly generating 50 dB(A) at a workstation, and a larger one generating 70 dB(A) at the same workstation. When all of them are operating simultaneously, the resulting sound level value will be 77 dB(A). If the major contributor is heavily treated to the extent it is no longer significant (meaning at least a 30 dB reduction, which will sure be costly!), the resulting sound level will still be 76 dB(A).

Note: This also has a consequence when attempting noise reduction. If you do not work out all the contributions from noise sources, you will have trouble making any progress. In this particular case, most people will acknowledge the highest contributor as 20 dB over the other noise sources will be easily noticed. However, when eventually adding all the smaller contributions, they turn out to be overwhelming to such an extent that treating the noisiest piece of equipment will not bring any significant numeric improvement.

2.3.6.3 Equivalent Sound Level

Let's consider a situation where a piece of equipment is generating noise. There are three phases of operation with respective durations of 5, 10, and 45 min and corresponding sound level values of 90, 80, and 70 dB(A). The equivalent sound level over 1 h will be 80 dB(A).

Let's now consider a situation where several pieces of equipment are generating noise at a given workstation. There are two phases of operation with respective durations of 5 and 15 min and corresponding sound level values of 100 and 80 dB(A). The equivalent sound level over 1 h will be 94 dB(A). Let's now suppose that due to noise control reduction measures, the 80 dB(A) value over 15 min has now been reduced to 50 dB(A) over 15 min. The equivalent sound level over 1 h will still be 94 dB(A).

Note: This has a consequence when attempting noise reduction. There will not be any improvement as long as the noisiest contributor has not been properly treated.

2.4 NOTIONS OF PROPAGATION

2.4.1 Propagation in Free Field

As implied by its name, an acoustic free field is a situation where there are no obstacles close to the sound source, the receiver, and the sound path. Under such conditions the sound waves from an omnidirectional point sound source will be spherical, and the sound pressure level L_p as a function of the sound power level L_w of the source and the distance d from source to receptor will be expressed as

$$L_p = L_w + 10 \log(Q/4 \pi d^2) - D$$

where Q is the directivity of the source:

Q = 1 for an omnidirectional source
Q = 2 for an omnidirectional source over a reflective plane
Q = 4 for an omnidirectional source in a corner

D is the extra attenuation (e.g., due to barriers, ground effects, etc.).

As other kind of sources can be obtained by adding sets of point sources over a given line or surface, one may note that the spatial sound level decrease with the distance varies as:

20 log(distance) for a point source
10 log(distance) for a line source of length L, as long as $d < L$
10 log(distance) for a surface source of maximum dimension L, as long as $d < L$

When dealing with a large source, it is always possible to divide it into smaller elements of area S whose dimensions will be such that the distance from the source to the point of calculation will be much larger than square(S) in order to try to assimilate it to a point source.

Further away, the spatial sound level decrease with the distance varies as 20 log(distance).

2.4.2 Propagation with Surfaces along the Way

The presence of surfaces along the propagation path will of course affect sound propagation due to reflective effects, or due to the screening effect should the surface of interest intercept the line between the sound source and receiver (cf. Section 2.4.5). On a reflective floor (note that this typically is the case of most workshops) propagation will steadily be reinforced [15]. In order to take into account those surfaces, it may be wise to use a computer prediction model [16–18].

2.4.3 Propagation over Terrain

Terrain will add a ground effect; more to the point, it may complicate matters through either screening or reflecting effects.

The National Research Council of Canada has developed a simple prediction model to study the noise impact from such sources as transportation corridors. The simple formula for normal grass terrain uses an extra attenuation term function of the distance d from the source to the receiver [19]:

D = 3.5 log (distance)

ISO 9613 has given for that purpose an attenuation term:

$D = 4.8 - (2\, h_m/d)\, (17 + 300/d)$

where h_m is the average height aboveground of the direct sound propagation path, in m.

2.4.4 Propagation in Rooms

Three cases of spaces may be considered [20] for a given sound excitation:

- Very small space (i.e., all dimensions are small with regards to the wavelength)

- Normal space (mean free path is large with regards to the wavelength; there is a practically constant reverberant field)
- Very large space (there is a spatial sound level decay)

In the first case, there usually are stationary waves.

In the second case, the sound pressure level L_p can be expressed as a function of the sound power level L_w of the source and the distance d from the source to the receptor as

$$L_p = L_w + 10 \log [4 (1 - \alpha)/A + Q/(4 \pi d^2)]$$

where Q is the directivity of the source, α is the mean absorption coefficient, and A is the total absorption (in m²).

In the third case, while there usually are some empirical models [21], it may be wise to use a serious computer model—associated with somebody who knows how to operate it.

2.4.5 Noise Barriers

A noise barrier is defined as any obstacle, natural or artificial, intercepting the line between the source and the receiver.

2.4.5.1 Noise Barriers Outdoors

Efficiency primarily depends on the path difference between direct and refracted sound on each edge of the barrier (which bluntly translates as the larger and higher, the better). Formulas to compute the efficiency of a noise barrier are given in [19, 22]. One may care to note that the efficiency of a noise barrier can be seriously hindered by the presence of reflective surfaces nearby (e.g., a wall or another barrier), as displayed in Figure 2.1. For this reason, it can be desirable to have absorptive treatment applied on the barrier on the source side.

The closer to the barrier the source and the receiver are, the more efficient the barrier effect will be. Standard ISO 9613-2 [23] limits the maximum gain in terms of noise reduction to 25 dB for a single barrier and 25 dB for multiple barriers outdoors.

2.4.5.2 Noise Barriers Indoors

While the physical phenomena involved are the same as above, the noise reduction efficiency of barriers indoors is seriously hampered by the presence of reflective surfaces [24].

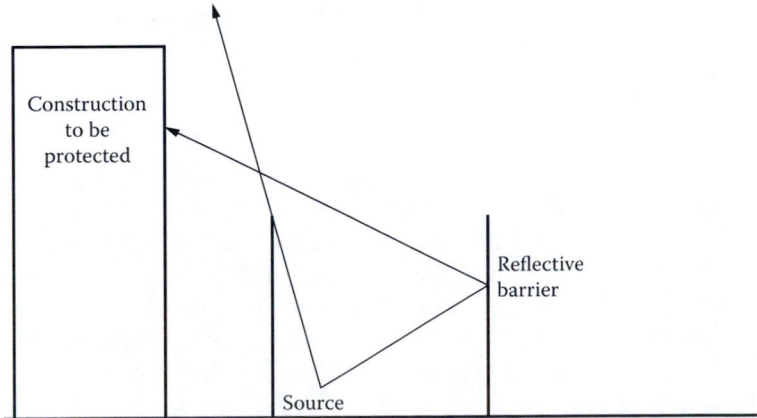

Figure 2.1 **Example of reflection on a barrier.**

The guidelines from standard ISO 17624 [25] limit the potential gain from a noise barrier indoors to 5 dB(A).

2.4.6 Examples

2.4.6.1 The Longest Propagation?

When the Krakatoa volcano erupted and exploded in 1883, the sound was actually heard a thousand miles away. Regarding infrasound, the resulting barometric depression circled the earth three times.

2.4.6.2 A Long Propagation

In 2001 the AZF chemical plant in Toulouse, France, blew up. Acousticians measuring plane noise at takeoff close to the Blagnac airport runway less than 10 km away saw their measuring apparatus overloading, while people in Villefranche, 30 km away, came out of their home thinking they had heard two trucks violently crashing into each other.

2.4.6.3 Propagation in a Tube

At the end of the 19th century two French physicists decided to make an experiment regarding the speed of sound as a function of frequency. A 2900 m pipe of 3 m diameter was built for that purpose. A large brass instrument tuba was used to generate a loud pulse. It was found that the resulting sound was heard during two returns. While no speed difference was noted (we now know that the speed of sound is not frequency dependent, don't we?), it was observed that the musical sound was gradually stripped of its higher-frequency contents to the extent that on the last passage, only the fundamental was heard (well, we also know by now that sound attenuation is dependent on the frequency to the power 2).

2.4.6.4 Propagation along a Curved Surface

Propagation along a curved surface (e.g., a wall or a vaulted ceiling) has been known since the Renaissance [6]. In Paris, one can experience it in most of the older underground stations due to their vaulted ceilings: When sitting on the bench of a platform, it is easily possible to understand the normal voice discussion happening on the bench of the opposite platform.

2.5 BASICS OF SOUND INSULATION AND ATTENUATION

2.5.1 Sound Attenuation

2.5.1.1 Sound Attenuation

One has certainly experienced sound attenuation within a building. For example, one may have been waiting in a public open space and discovered that the discussions held in various parts of this open space were heard, if not even understood, while others were not, either because they were too far away or because they were masked by the background noise.

Let us make a distinction: We will talk of *attenuation* when the sound source and the receiver are contained within the same volume. This attenuation will then simply be expressed as the difference between the sound pressure level generated at the source and the sound pressure level measured at the receiver.

Incidentally, a particular case of volume is the exterior space. One can also talk of attenuation there (cf. Sections 2.4.1 to 2.4.3).

2.5.1.2 Examples

2.5.1.2.1 Vegetation Outdoors

Vegetation is supposed to bring extra attenuation when located along the path of propagation outdoors. When asked about its effect, most residents enjoying the view of vegetation hiding a transportation corridor or some machinery will eagerly remark that it reduces the noise annoyance (though similar tests performed under laboratory conditions may yield a somewhat mitigated result [26]).

However, physical studies [28] show that the additional noise reduction brought by a curtain of trees is 0 ± 3 dB in the frequency range of interest! The possible adverse effect comes from reflections on branches and leaves.

2.5.1.2.2 Absorptive Treatment in a Room

As sound propagation in a room is highly dependent on reflections on the inner envelope, absorptive treatment is an essential part of noise control measures. Experiments have shown that an absorptive treatment can easily reduce the noise levels in a workshop by at least 3 dB(A) due to the elimination of reflections propagating sound energy throughout the workshop.

2.5.2 Experiencing Sound Insulation

One has certainly experienced sound insulation within a building. For example, one may have been annoyed by the discussions held in the office next door, or one may have been puzzled to hear people talking at night in the flat next door.

When the sound source and the receiver are located in different rooms, we will talk of insulation. The sound level difference will be defined as the sound level in the source room minus the sound pressure level in the receiving room.

One will, however, find that with the same construction technique applied, the resulting sound insulation is not similarly perceived, depending on the dimensions of the room, the nature of its cladding, and the nature of the walls and floors around. This means that a specific quantity must be defined to characterize the wall or the floor independently of its environment: this is known as the sound reduction index that can be measured in the sound insulation test suite of an acoustic laboratory.

2.5.3 Sound Reduction Index

2.5.3.1 Definition

The sound reduction index is the acoustic characteristic of a wall or floor for sound reduction purposes. It is given the symbol R, and it is not dependent on the characteristics of the room and its envelope. It is determined through laboratory measurements (though under specific circumstances it can be computed with a limited accuracy). The measurement procedure is given in Section 3.3.2.

2.5.3.2 Predictive Computation

Computation of the sound reduction index of a single wall can be carried out using a prediction model. The crudest of these models is known as *mass law*; as implied by its name, it relies on the mass per surface unit of the wall, noted m. The main term of this law is $20 \log(mf)$, meaning that doubling the mass will result in a 6 dB gain. Also, doubling the frequency will result in a 6 dB increase. Such a model can be enhanced by introducing a stiffness term.

Computation of the sound reduction index of a multiple wall can also be carried out using a prediction model. The crudest of these models features a description of each layer as per above, plus terms pertaining to the propagation inside the gap (filled by either air or a material) and to the coupling between the layers. Also, doubling the frequency will typically result in a 12 dB increase for a double wall.

A good insight can be gained in [29], which considers a one-dimensional model of single or multiple walls.

It is also possible to perform this computation through dedicated prediction computer programs. Of course, the more powerful the program, the trickier it is for somebody not familiar with the physics of sound transmission.

As implied by its name, a model is only a representation of reality. One must always try to master it first through the modeling of a well-known case to find out how to tune it. More to the point, one may not expect the model to always give an accurate prediction, but one may assess the probable variation of the sound reduction index due to a change in composition or dimensions.

2.5.3.3 Examples

2.5.3.3.1 Sound Reduction Index of a Few Partitions

Here are a few examples of R_w or Sound Transmission Class (STC) values:

20 cm thick reinforced concrete wall: 59 dB
10 cm concrete block wall: 45 dB
Two 13 mm plasterboards screwed on the same side of studs: 29 dB
Plasterboard partition made of two leaves of 13 mm plasterboards screwed on each side of 48 mm studs, with mineral wool in between: 47 dB
Plasterboard partition made of two leaves of 13 mm plasterboards screwed on each side of 48 mm studs, without mineral wool in between: 42 dB

2.5.3.3.2 Sound Reduction Index of Concrete and Mineral Wool by a Simple Model

TNO (Nederlandse Organisatie voor Toegepast Natuurwetenschappelijk Oderzoek) (Netherlands Organization for Applied Scientific Research) developed a quite simple and easy-to-use prediction software for the assessment of the sound reduction index of a single or double wall [30]. This software worked rather well as long as the user understood some basic physical transmission phenomena in order to be able to simplify and fit the reality into the model. Unfortunately, adding a mineral wool on one side of a 20 cm thick concrete wall would boost the sound reduction index from 50 to 70 dB. This was due to the absence of safeties on the entries (who would put unprotected wool directly onto a wall?).

Lesson Learned: Do not blindly trust a model.

2.5.4 Sound Insulation

2.5.4.1 Definition

Sound level difference is defined as the difference between the sound pressure level $L_{pemitted}$ in the room where the sound source is located (emission room or source room) and the sound pressure level in the receiving room $L_{preceived}$:

$$D_{nT} = L_{pemitted} - L_{preceived}$$

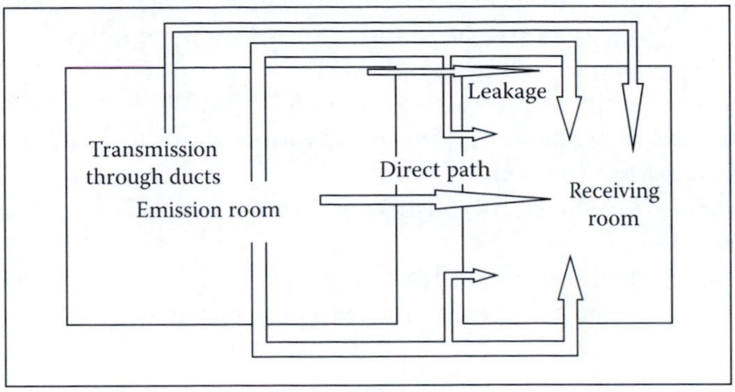

Figure 2.2 Acoustic transmission paths between spaces.

Standardized sound insulation is defined as the sound level difference to which a weighting term taking into account the reverberation time T of the room, as compared to a reference reverberation time T_0, is added. This usually is the quantity that is specified in requirements books and regulations.

$$D_{nT} = L_{pemitted} - L_{preceived} + 10 \log(T/T_0)$$

2.5.4.2 Computation

As compared to a laboratory situation where sound transmission solely occurs through the separating wall, sound transmission between two rooms of a building can be decomposed into:

- Direct transmission.
- Flanking transmissions through the various constructive elements linked to the separating wall. Such flanking uses coupling coefficients k_{ij} that are defined in the standards of the ISO 12354 series.
- Leakage around some construction elements.
- Secondary transmissions by means of duct layouts.

These are displayed in Figure 2.2.

The basic idea is for a computation model to take into account all of these contributions. Some commercial computer software offer the possibility to use a readily available library of components (complete with the k_{ij} values). Whenever a new material must be introduced by the user, it is wise to do so by modifying an as similar as possible material.

Most of the time, such software is able to produce graphic results (e.g., global sound insulation between the two rooms of interest, plus sound insulation through each main path). This helps identify which component needs urgent improvement.

2.5.4.3 Which Insulation for Which Use?

Here are a few sound insulation ranges of sound insulation values that can normally be experienced:

- Between offices with removable partitions: 25–35 dB. (Note: 30 dB is considered by several standards to be the minimum sound insulation value required to allow two of the same kind of activities to be carried out in two adjacent rooms. While a 25 dB sound insulation value can still be found, typically in front of wall elements, including doors, it is not recommended by standards.)

- Between a doctor's office and waiting room: 45–50 dB. (Note: 45 dB is considered by several standards to be the minimum sound insulation value required to allow privacy between two adjacent rooms.)
- Between modern dwellings: 50–55 dB. (Note: This usually is the range required by most regulations.)
- Between cinema theatres: 65–70 dB. (Note: This usually is the range required by the background noise levels targeted by the standards of the industry.)

2.5.5 Examples

2.5.5.1 Waiting Room

A doctor's consulting office featured walls made of 20 cm thick plain concrete blocks with a painted concrete finish. This had a sound reduction index value of 61 dB, which theoretically permitted a sound insulation value of at least 51 dB to the waiting room next door.

Unfortunately, there was a single door in this partition. In order to make its opening easy by restricted mobility people or elderly people, it was made sure that it would not be too hard to open or close it. This resulted in not sufficiently airtight seals, with a resultant sound insulation value between the doctor's office and waiting room slightly above 40 dB: This meant some of the discussions held in the office could be heard, and even occasionally understood, in the waiting room. In order to solve the problem, it was necessary to build air lock-mounted doors (i.e., two set of doors with a small lobby in between), as a higher-performance single door would have been too heavy to operate by some patients. And due to the space required to be able to turn a wheelchair or a stretcher, a sizable part of the area had to be redesigned and refitted.

Lesson Learned: Always take into account acoustics requirements as well as safety/accessibility requirements.

2.5.5.2 Improvement of the Façade of Dwellings and More

On a regular basis the acoustician is required to make a submission for a study on the improvement of the façade sound insulation of condominiums. Most of the time the offer to perform a diagnosis of the sound insulation between flats, while assessing the performance of the existing façade, is turned down. And usually 2 years later those clients come back, requiring a new submission, as they are now complaining of annoyance between flats much more than before.

The reader probably guessed it: The intrusion of noise from the exterior used to cover whatever noise came from the flats around. When the façade was upgraded, there was less masking noise from the exterior, and the intrusion of noise from other parts of the building became more noticeable.

Lesson Learned: Always make an inventory of all the noise sources before attempting to set acoustic objectives.

2.5.5.3 Envelope Made of an Absorptive Material

In the 1970s the French Ministry of Culture launched a concept named "zenith" to try to reduce the cost of pop concert facilities. The original concept, by architect Chaix & Morel, called for a thick, absorptive material between two canvas layers that could be mounted like a circus tent.

Of course, while it provided significant sound reduction in the high-frequency range, it did not provide any in the low-frequency range. The concept was later evolved into a heavier (and more permanent) complex.

2.5.5.4 Designing, Building, and Commissioning

A building project was initially designed using 20 cm thick concrete walls. While this was accepted at the design stage, the structural engineer admitted that 20 cm thick plain concrete blocks with a concrete finish could also be used. Next, the client found it more expedient to use hollow concrete blocks of the same thickness, as they were cheaper and structurally acceptable. However, the sound reduction index was now down by more than 5 dB. Eventually, the contractor also managed a simplification of his own by replacing the concrete finish by a glued plasterboard, which resulted in a further 4 dB missing due to the resonance frequency introduced by the small gap between the plasterboard and concrete blocks and the unchecked porosity of the concrete blocks.

Lesson Learned: Always check with the acoustician before substituting a material for another.

2.6 BASICS OF REVERBERATION TIME

2.6.1 Absorption

2.6.1.1 Experiencing Reverberation (or Its Lack)

On entering a premises, one has often experienced an acoustic feeling of the place (e.g., it sounds lively, or alternatively, it sounds dead). Of course, it is tempting to quantify this impression. The corresponding quantity is known as *reverberation time*. In the olden days, it was measured by emitting a loud noise while pressing the chronometer and stopping it when it was no longer audible.

A professor at the school of architecture eventually discovered that the lecture theatre he was using was more or less dead according to the number of cushions that would be brought in by his students. This eventually led him to the notion of equivalent absorption area.

2.6.1.2 Definition

The absorption coefficient of a material is represented by the symbol α. It is defined as the ratio of absorbed energy over incident energy. One can see it cannot be physically greater than 1. However, there are some test reports that will give a value greater than 1 due to the standardized procedure (cf. Section 2.11).

The equivalent absorption area A of a material is defined as the product of its area S by its absorption coefficient α:

$$A = S \, \alpha$$

2.6.1.3 Absorptive Materials

Absorptive materials are materials whose absorption coefficient is at least 0.5. Those are typically made of porous or fibrous materials.

One should keep in mind a few physical facts. To start with, the thicker the material, the better its absorptive performance in the low-frequency range will be. Next, an absorptive material features an airflow resistivity, and one of the quickest ways on site to find out whether a so-called absorptive material really is up to the task is to try to blow through it. If this proves impossible, then chances are high that it is not absorptive.

Now let's look at a material exposed to an incident sound wave. Part of the energy will be reflected, while another part will be transmitted through it and a last part will be absorbed, as displayed in Figure 2.4.

More to the point, while the sum of the squares of the coefficients will be equal to 1, the individual components will vary according to frequency. For example, a 10 cm thick mineral wool will be an efficient absorber in the mid-frequency range. In the low-frequency range, it will perform poorly as an absorber; however, in the low-frequency range it will let sound energy through.

This has an important consequence when dealing with the internal acoustics of a room: while glazing definitely is not absorptive as per the above definition, the amount of sound energy transmitted through it in the low-frequency range will be such that it will be equivalent to a low-frequency absorber.

2.6.2 Reverberation

2.6.2.1 Experiencing Reverberation (or Its Lack)

On entering a premises, one has often experienced an acoustic feeling of the place (e.g., it sounds lively, or alternatively, it sounds dead). Of course, it is tempting to quantify this impression. The corresponding quantity is known as reverberation time. In the olden days, it was measured by emitting a loud noise while pressing the chronometer and stopping it when it was no longer audible.

Basically, reverberation comes from the multiple reflections on the inner envelope of the room, as displayed in the example of Figure 2.5. In a real hall the reverberation time curve will show a succession of waves, as illustrated in Figure 2.3.

2.6.2.2 Reverberation and Echoes

Reverberation is made of numerous acoustic reflections that are spread over time so as to avoid any gap longer than 100 ms and preferably no more than 22 ms from any other similar reflection.

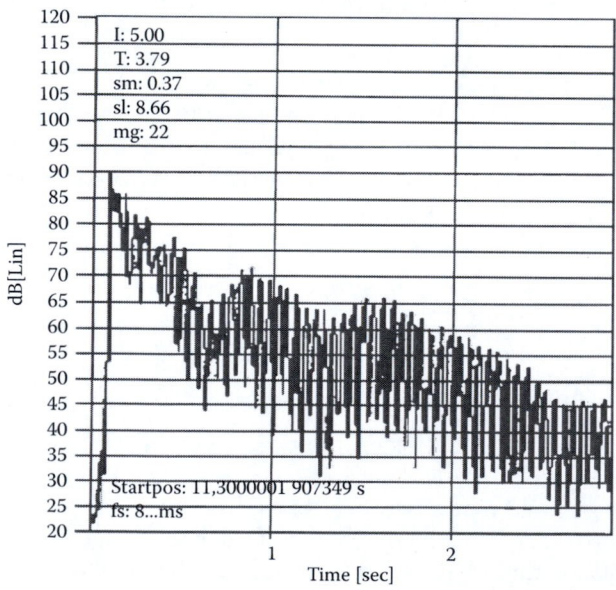

Figure 2.3 Example of reverberation time curve in the 500 Hz third octave band.

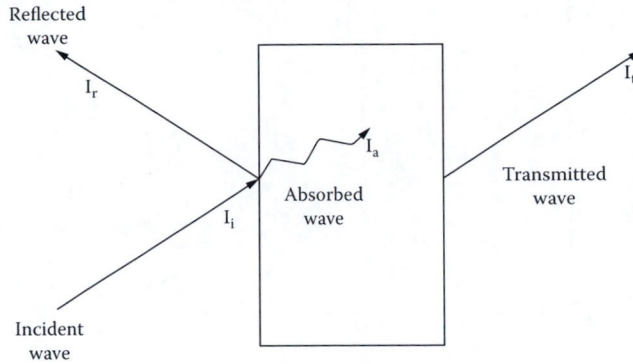

Figure 2.4 Incident, reflected, absorbed, and transmitted waves.

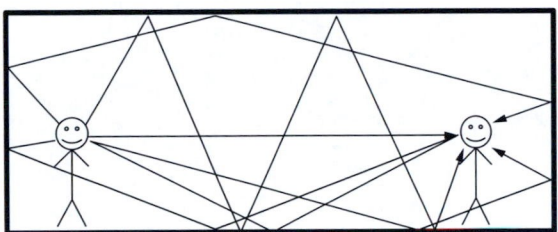

Figure 2.5 Example of acoustic reflections in a room resulting in reverberation.

Echoes are the product of multiple reflections arriving within a very short time span (no longer than 22 ms) well after the direct sound (i.e., more than 100 ms).

Reverberation time is a bit similar to the time constant of a physical system: the longer the time value, the greater the stability, but on the other hand, the slower the reaction to a change of excitation. This has an important consequence: In a reverberant room one will not have trouble being heard (a whisper would do the trick, as the sound is well sustained), but one will have trouble being understood due to the bad rendition of consonants.

2.6.2.3 Definition of RT

Reverberation time (RT) is defined as the time span needed for the sound level to decrease by 60 dB from the moment the sound source is stopped. This is also written as RT_{60}.

Looking at a typical reverberation curve, one can note that there typically are three regions, as displayed in Figure 2.7.

It can be of interest to assess the early reverberation time (early decay time (EDT)), which is assessed over the first 10 dB decrease (and sometimes written as RT_{10}). This will be discussed in Chapter 6.

By the way, looking at a typical reverberation curve, one can sometimes note a change of slope appearing in the decrease, as displayed in Figure 2.6. This usually indicates the presence of a flutter echo.

2.6.2.4 Simple Computation of RT

A simple predictive assessment of the reverberation time can be carried out from such basic quantities as the volume and absorptive areas of the room. The famed Sabine formula is

$$RT = K \ V/A$$

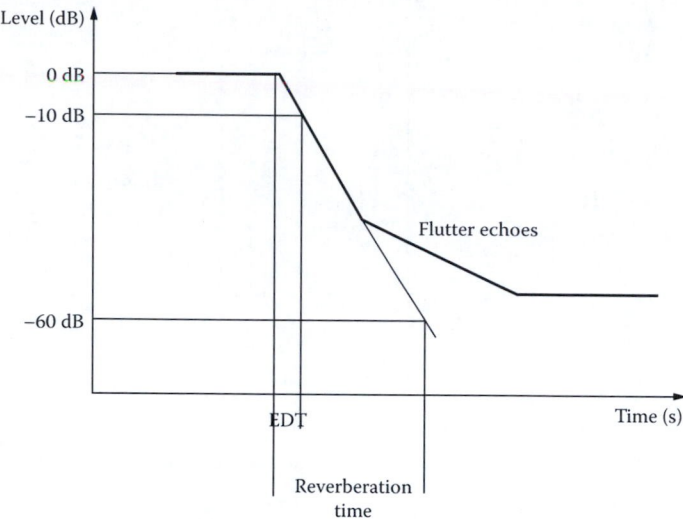

Figure 2.6 Example of stylized reverberation time curve.

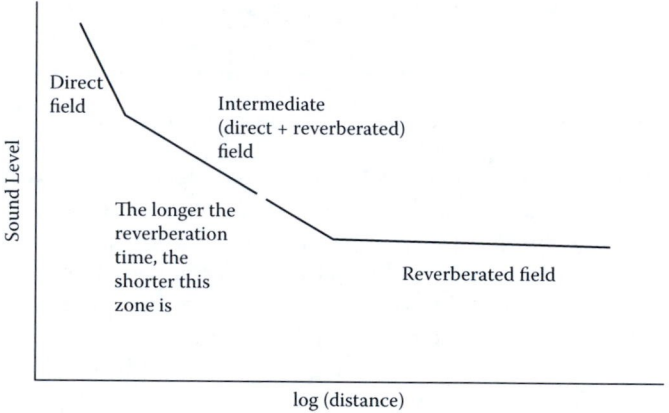

Figure 2.7 Spatial sound level decay.

where *V* is the volume of the room, *A* is the equivalent absorptive area, and *K* is a coefficient with a value of 0.163 in metric units.

While simple, this formula is actually supposed to be valid for volumes with rather reflective surfaces and similar dimensions. The greater the absorption, the smaller the accuracy of the prediction is. Nevertheless, it does give interesting information regarding tendencies (e.g., increasing the volume, or adding extra absorption in a room).

To try to adapt to more current room volume shapes or partial absorptive treatment, similar formulations have been developed by Eyring and Millington. But on the whole, the Sabine formula is best suited to most situations [31].

2.6.2.5 Which RT for Which Use?

The reverberation time is akin to the time constant of a physical system: The longer the time, the slower the reaction speed and the better the stability. One may deduce that for harmonic music (i.e., music where chords are the primary expression, such as Gregorian singing), a long reverberation time is required (e.g., over 3 s), while for melodic music, where each note counts, a shorter reverberation time (e.g., 1.6 s) is needed. For speech and singing, where the text is of importance, an even shorter RT value is looked for (e.g., 1.1 s). These aspects are developed in Chapter 6.

The reverberation time may also indicate the eventual lack of acoustic absorption in the space. As less absorption means a higher sound level, one must keep in mind that for noise control purposes, long reverberation times should be avoided (e.g., in a dining room).

2.6.3 Spatial Sound Level Decay

While widely used, the reverberation time is not always suitable as a descriptor of internal acoustics; e.g., in a flat encumbered space it will feature a rather low value, sometimes similar to that for a small lounge, while people inside will find the acoustics uncomfortable. For such situations, one may use the spatial sound level decay DL2, which is the rate of sound level decay with doubling of distance [32]. Figure 2.7 displays the stylized curve of decrease versus distance, on which it is possible to distinguish the direct field region close to the source (where the influence of the room is not yet felt) and the far field region (where one is left with the reverberant field). In between stands the intermediate region, which is controlled by the acoustic treatment of the room.

2.6.4 Examples

2.6.4.1 RT and Music

Gregorian singing can be heard in various cathedrals and churches where the RT value is well over 3 s. At the opposite side of the scale, the Scala of Milano Opera features a 1.1 s RT to enable the singers of Bel Canto to display their skills of articulation.

2.6.4.2 Bad Reverberation and Speech

When the Fogg Art Museum of Boston was designed, its architect chose for the conference auditorium a volume that had similar dimensions in the three directions; to top it all, it also featured a cupola. As absorptive materials were a bit too scarce for comfort, this made for an impressive diffuse field with some focusing too.

The acoustic result was so catastrophic that the trustees decided on affixing a commemorative plate at its entrance explaining why this auditorium had been christened with the name of its architect.

2.6.4.3 Propagation between a Confessional and a Crypt

A confessional was built on the side and at the end of a crypt. Soon the priest in charge was puzzled to be told that most of the proceedings were fully understandable! It turned out that should one stand at a point symmetric to the confessional, there was a low enough background noise level and strong enough focusing to enable prying ears to understand every word.

2.6.4.4 Absorptive Turned Reflective

In order to be able to use a small room initially intended for storage as an extra classroom, a school had an absorptive treatment made of perforated plasterboard with mineral wool applied on the ceiling.

A few years later the new director decided a refurbishment was in order and a painter duly sprayed the ceiling. As the ceiling material got clogged with paint, it lost all its absorptive properties.

Lesson Learned: Painting an absorptive material will usually result in the loss of its properties due to the closure of the porosities.

2.6.4.5 Resonances of Volumes

One has probably experienced some situation in which a room seems to be more responsive in some frequency bands than in others. But is it possible to compensate such a behavior using extra electroacoustic power? Here is a small example to illustrate how hard it might be on such a small volume as a brass instrument.

During experiments at IRCAM in the 1980s, recordings were made inside the mouthpiece of a brass instrument. When emitted in the mouthpiece again using a tape player and a small loudspeaker, the noise from the artificially excited instrument sounded like the real one. Of course, it was then very tempting to change the frequency and make the instrument "play" a scale or even a tune. Some colleagues from the computer department tried it; they found that it was difficult to play evenly due to the resonances of the instrument. They then devised a simple program to adjust the gain for each note on the scale. When a full scale was played up and down, there was a nasty smell, and the molten remains of what was left of the loudspeaker when it fell on the table.

Lesson Learned: It usually is not worth attempting to excite a volume, even a small one, between resonance frequencies.

2.7 NOTIONS OF INTELLIGIBILITY

2.7.1 Definition

Intelligibility is the ability to understand speech under given conditions.

Of course, while there are situations where one wants good intelligibility (e.g., for the listener of a speaker in a conference hall or for a traveler attempting to listen to the PA system), there also are situations where one does not want to be disturbed by undue speech (e.g., in a open-plan office); there are even situations where one wants to make sure speech is not understood (e.g., in banks).

Several levels of understanding can be identified:

- Discretion is a situation where a sizable part of the speech signal is not understood by a listener, thus limiting the annoyance created by the speaker. This is the best one can achieve in an open space.
- Privacy is a situation where speech is not understood under normal circumstances. This implies partitions (open space is ruled out).
- Security is a situation where speech is not understood under any circumstance.

2.7.2 Factors

Various factors can influence the intelligibility of spoken messages. Here are a few of them:

- Signal-to-noise ratio: The louder the background noise, the smaller the speech signal will emerge above it and the more difficult it will be for the listener.
- Reverberation time: The longer the reverberation time, the more difficult it will be for the listener to catch the consonants and understand the spoken words.
- Distance between the speaker and the listener: The longer the distance, the more the acoustic attenuation, and thus the lower the received speech signal.
- Visual cues: The better the vision of the speaker by the listener, the better his understanding of the speech signal.

2.7.3 Assessing Intelligibility

The assessment of speech intelligibility can be performed through a list of words or sentences being read by a normal speaker to a panel of normal-hearing listeners. Unfortunately, this method calls for a significant number of participants, which makes it costly and unpractical. More to the point, it is quite time-consuming.

Depending on the native language of the speaker and the listeners, scores can vary by as much as 20%.

Other methods call for an artificial test signal being diffused at the speaker location and measured at various listeners' positions. The measurement results can be expressed using some intelligibility descriptors such as the speech transmission index (STI) (cf. Section 6.5). The resulting score is not language dependent.

2.7.4 Which Intelligibility for Which Use?

Depending on the nature of the speaker–listener relation, various intelligibility scores can be aimed at. For example, in a theatre or a conference room, one will attempt to have as high an STI value as possible (at least 0.70, preferably 0.90), while in an open-plan office one will try to have an STI value well under the 0.50 mark between noncollaborating workstations (cf. Chapter 7). The corresponding results will usually be achieved through a mixture of background noise level adjustments, and acoustic propagation enhancement or prevention.

2.8 NOTIONS OF ANNOYANCE AND DISTURBANCE

So far one had contentedly dealt with physical quantities that can be measured. Unfortunately, more often than not, the acoustic receiver happens to be an ear and its complicated brain processing behind it, and there is no measuring instrument to read a possible annoyance or disturbance level. More to the point, are the usual noise metrics (i.e., relying on the ubiquitous A weighting) really adapted to provide physical quantities correlated to annoyance [33, 34]?

2.8.1 Noise Annoyance

Noise annoyance can be defined as "a feeling of resentment, displeasure, discomfort, dissatisfaction or offence which occurs when noise interferes with someone's thoughts, feelings or activities" [35, 36].

Calling on feelings means that quite a number of surveys will be necessary to try to apprehend the applicable criteria [37–39, 49] and investigate the global situation [40, 41]. More to the point, annoyance seldom results from one specific sound source, and there often are other factors, auditory or visual [42], as well as cultural [48], that will influence the score. The case of airports has of course proven interesting [43] due to the high sound levels involved. But traffic noise also provides interesting cases due to its universal location [44] and time variability.

2.8.2 Noise Disturbance

Noise disturbance can be defined as "any noise, sound or signal which unreasonably disturbs the comfort, peace, or repose of another person" [45]. With such a definition lawyers are looking forward to happy exciting moments attempting to define what "unreasonably" actually represents! Therefore, a legislator eventually came up with a definition of his own [46]:

> Noise disturbance is defined as any source of sound, which exceeds the noise limitations, permitted by the noise ordinance. Sources of sound shall include but not be limited to the following: amplified music, loudspeakers, radios, televisions, stereos, musical instruments, powered toys or models, swimming pools or spas, industrial machinery, manufacturing equipment, pole drivers, air compressors, paint sprayers, motors, pumps, blowers, air conditioners, cooling towers, ventilation fans, fork lifts, loaders, tractors, animals, concerts, mechanical equipment, human voices, electrical appliances, vacuum cleaners, powered equipment, chain saws, motor vehicles and attached equipment not operated on a street or highway, etc.

2.8.3 Statutory Noise Nuisance

In order to be able to apply noise control bylaws, a legislator eventually came up with a notion of statutory noise nuisance [47]:

> A statutory noise nuisance is more than where the noise is a mere annoyance, but where it is viewed to have a significant impact on the health and well-being of those affected. Many factors are taken into account in determining if the noise amounts to a statutory nuisance, including:

- Location. Is the noise typical for the area? A cockerel crowing in the countryside would be more accepted than that in a quiet urban residential road [cf. Section 2.10.6.3].
- Time of day. A statutory nuisance can exist at anytime of the day, however, the effects of noise late at night when most people are sleeping would be given greater weight than the same noise occurring during the day.
- Frequency. How frequently is the plaintiff affected by the noise? Clearly noisy parties every weekend would be viewed differently to one held occasionally.
- Duration. How long is the plaintiff disturbed? A dog barking at the postman or barking at cats occasionally passing by would be viewed differently to one barking most of the day.
- Intensity. How loud is the noise? How intrusive? We all have different noise thresholds and tolerances. In determining nuisance the judgement would be how the noise would affect an ordinary individual, not someone who had a particular sensitivity to the noise complained of.
- Number of people affected. A view will be taken on the number of people who are, or could be, affected by the noise. If only one person complains when the whole street could equally be affected, then there could be a challenge that the individual making the complaint could be unduly sensitive.

A survey of noise control legislation and regulations has been published by I-INCE [50].

2.8.4 Noise Annoyance and Noise-Induced Disturbances

Noise-induced disturbances are not just some idle talk subject. The social costs of traffic noise in the European Union (EU) have been found to be €40 billion per year [51].

Exposure to unwanted noise may result in noise-induced disturbances, the most commonly thought of being sleep disturbance, but some other effects (e.g., cardiac problems) can also occur.

In order to try to predict the potential effects on health from the construction or extension of transportation infrastructures or large public or industrial equipment, several research programs have been launched, with the ultimate aim to correlate noise and vibration physical values to annoyance and related disturbances. One of the first tasks has been to try to find out whether suitable noise and vibration indicators are currently available. An inventory has been made (e.g., [34]), and the next step will probably be to confront them with existing situations before attempting to create a predictive model.

Noise, but also ground vibrations and a number of events (e.g., trains passing by), has been found influential [52].

Sleep disturbance is a frequent complaint [53] in noise-exposed situations, leading to an increase of blood pressure for noise level values L_{Amax} inside the bedroom sometimes as low as 35 dB(A), and awakening occurring for 42 dB(A) [53]. Problems are not limited to the bedroom, as physiological effects have been reported on workers exposed over 55 dB(A) [54].

2.8.5 Emergence, Exceedance, Intrusiveness

Typical community noise control regulations will typically state noise limits. Now while a simple sound level value (e.g., a L_{Aeq} measured at the plaintiff's location) may look simple enough to assess and use, it will not always correlate with the annoyance or to all situations (e.g., if the background noise level is high enough, there probably will not be much of a problem, as the intruding noise will probably be masked). But if the background noise level is rather low, then the intruding noise will be perceived and identified.

To circumvent this difficulty, some standards and regulations use a quantity that describes the excess of particular noise over the background noise level. This quantity is defined as the difference between the ambient noise level when the equipment is operating and the background noise level measured when it is not. It is called *emergence* in the recent (2014) FDIS ISO 1996 [55] and can also be found as *exceedance* in the United States or *intrusiveness* in Australia. For example, in France a recommendation by the Health Ministry states that a noise from an equipment or activity can be considered potentially annoying when its emergence value is greater than 5 dB(A) in daytime and 3 dB(A) in nighttime [56]. The local authorities can give precise information regarding those restrictions.

2.8.6 Examples

Examples of how a noisy situation can pester one's life to the point of action are rife. Here are just a few.

2.8.6.1 Low-Frequency Noise in a Dwelling

The director of a public service called the acoustician: He could not live in his service apartment due to a low-frequency noise that appeared on a rather regular basis. On visiting him in the morning, the acoustician could not hear anything untoward. On performing careful measurements late in the evening, he eventually found an emergence of less than 0.5 dB(A), with a predominant contribution in the 31.5 Hz octave band. The director's ear was really tuned for the detection of that noise. It was found that this noise came from the environment rather than from the next building, but it was not possible to investigate further.

A couple of years later it was eventually found that this noise was radiated by the elevated urban railway viaduct some 300 m further.

Lesson Learned: Some sounds may look harmless to some people and unbearable to others. More to the point, attempts to identify the sound source of interest can take a lot of time.

2.8.6.2 Barking Noise and Other Animal Noise

Back home, a man remarked to his friend that the dog next door was really pestering everybody with its frequent barking. Having him repeat that, the friend quietly pointed out that the dog on the other side of the house was considerably noisier, but apparently not perceived as such due to its being his friendly pet!

A couple who had led an urban life chose to retire in the countryside. Soon they felt annoyed by the cock and went for a court action. It took all the might of the judge to explain that the noise from a cock was to be considered normal in the countryside!

Lesson Learned: According to one's customs and inclinations, a noise may be perceived as annoying or not—and considered as such.

2.8.6.3 Speech Noise Outside

A bus stop with a small shelter was installed in front of a house in a small village. The dwellers were soon fed up with the youngsters meeting up there late in the evening. After petitioning the mayor for either the removal of the shelter or its relocation, the inhabitants took the matter in their own hands: they managed to saw off the shelter and had it implemented a few hundred meters further!

Lesson Learned: Moving the source away will reduce the noise and inconvenience (but an authorization usually is required!).

2.8.6.4 Noise Reduction at the Source (Never to be Imitated)

A man irritated by the frequent noise of teens on the lawns downstairs opened his window and shot at them.

Lesson Learned: Exposure to unwanted noise may lead to disastrous moves. Prevention must be implemented whenever possible.

2.9 STANDARDIZATION IN ACOUSTICS

2.9.1 Purpose

As a general rule, standardization is meant to help promote exchanges between the various interested parties [57]. As an example, it may provide guidelines on using certain types of noise control equipment (e.g., noise barriers or enclosures) or provide basic guidance on fitting out spaces (e.g., a workshop or an office), or it may outline the methods to perform measurements. The basic idea is to try to inform all interested parties (including users) of the potential and the drawbacks of some noise control solutions or measurement methods. Unless specifically required by regulations, the application of standards is purely voluntary.

Regarding the latter, the measurement methods may be precision (laboratory measurements) or survey (field-grade measurements). In both cases the operating procedure (requirements on the measurement apparatus and the measurement place) is described, and the various parameters to be reported are clearly identified. Last but not least, the schedule of the measurement report is displayed, including the quantities to be reported.

In building acoustics there now is a trend to try to separate the measurement procedure proper and the preparation of the test sample. The former standard will be solely concerned with the measurement apparatus and method, while the later will solely deal with the mounting of the sample (e.g., which sealant to be used on the edges, how to store the sample prior to construction and testing, etc.).

Of course, the field of standardization is far from limited to building acoustics; it may concern such domains as building acoustics, occupational noise, environmental noise, physio-acoustics, metrology, etc.

2.9.2 Process

The elaboration of standards is rather well codified: First, one or several bodies will remark on the need to have a standard covering a specific item. Next, the request for such a standard (known in the trade as new work item proposal (NWIP)) will be submitted to the relevant standardization committee. If accepted, a convenor and a secretary will be nominated and invitations to participate will be sent. A workgroup will be created (or an existing workgroup will be tasked with this job). The members will first produce a draft (known as committee draft (CD)) that will be circulated by the representatives to their bodies of origin. When agreed upon, a draft standard (DS) will be officially circulated by the standardization body to all identified interested parties for comments. Those comments will be examined by the workgroup and either adopted or rejected. A revised version of the DS will then be circulated for voting. The whole process usually takes between 2 and 3 years. More to the point, standards are usually called to scrutiny after 3 to 5 years, taking into account the comments from the users. According to the severity of those comments, they may be either given another lease of life for the next 5 years, revised, or withdrawn.

Due to the different kind of parties involved (e.g., governmental entities, consulting engineers, producers, end users, etc.), the exchanges may sometimes be heated! If a compromise is reached, it may not reflect the state of the art due to the various accommodations needed to satisfy everybody (or at least not to offend anybody). There have been instances where no compromise was reached, yet the political pressure on editing a particular standard was such that several quite different methods found their way in the standard!

The standardization entity can be national (e.g., BS in Great Britain, AFNOR in France), European (EN), or international (ISO). In addition, some industries and engineering syndicates have their own set of standards. Quite often there now is a joint standardization process between ISO and EN that often results in a standard with appendixes that have informative status with ISO and normative status with EN. In Europe the national standards will simply translate the EN standard and include it in their list. Incidentally, one might care to remember that no national standard of a European country may be in opposition to a European standard, so there is quite a lot at stake there.

2.9.3 A Few Points to Beware Of

A test report is not limited to a curve displaying the measurement results. It also includes some important information, such as:

- The name of the entity performing the test. (Is it known and respected?)
- The name of the entity that required the tests (which probably has a stake in the measurement results).
- The description of the test facility. (Does it comply with the standard requirements regarding testing facilities?)
- The description of the measurement apparatus. (Is it regularly checked?)
- The description of the test sample. (Careful—Does it correspond to the product whose test report had been required? Do the masses and dimensions correspond to the actual specimen? When it reads "the upper plate has been laid on the mineral wool," does it mean there was no mechanical fixation?) Is this description labeled "according to the manufacturer"? In the affirmative, one should be worried!
- If applicable, the description of the operating conditions. (For example, for a shredder, what kind of material was used for the test? Does it correspond to normal practice? Are the measurement points clearly identified? For a HVAC unit, did it operate at the required operating point (temperature and pressure)?)
- Is the measurement uncertainty evaluated?
- Are the measurement results better than anticipated? In the affirmative, does the laboratory have a reputation for usually getting better values than the average other laboratories?

Let's be clear: While laboratories are not crooks, some of their clients may be not exactly straightforward. There are so many factors that can influence the results that one had better be careful. The above checklist is not exhaustive but will give room for some questions.

2.9.4 Examples

Here are a couple of examples regarding some test reports. They are all real (and of course, the interested parties will be nameless), and unfortunately, they are far from isolated incidents.

2.9.4.1 HVAC Individual Unit

A large hotel project featured individual HVAC units in each guestroom. The specification books had clearly indicated both a maximum sound level value in the room and a maximum sound power level for the equipment. Similarly, the HVAC engineer had stated thermal requirements on those units. Accordingly, the contractor was required to produce an acoustic test report of his equipment. Looking at the measurement results, it turned out that thermal and acoustic measurements had not been performed at the same laboratory. This rang a bell in the engineer's mind, and it was quickly found that the acoustic measurements had not been performed under the required thermal load. A new test performed simultaneously showed that the requirements were not simultaneously met.

2.9.4.2 HVAC Large Building Unit

A big building was to be fitted with a large set of air coils. Due to the size of the equipment, the manufacturer refused to send part of it to the laboratory and proposed to perform the tests at his own facility that complied with all relevant standards according to him. The acoustic engineer went to investigate and found that the test room intended for the individual fan had dimensions not complying with the standards; furthermore, the measurement apparatus had not been checked for 15 years, as testified by its documentation! When it came to the whole equipment, it quickly turned out that the fans were only operating at half speed.

2.9.4.3 Ceiling

When prompted to produce the laboratory test reports for absorption and for flanking normalized level difference, the representative of a big ceiling company used to hand in a so-called technical document that featured the copy of those test reports with the name of the product on the cover sheet. However, the description of the samples was unusually short. Looking closely at them, it would eventually turn out that two different types of ceiling featuring the same generic brand name had actually been tested in each case in order to display better results.

2.9.4.4 Door

A puzzled economist phoned an acoustician friend of his with a surprising question: Could a door without doorstep and seals reach a sound reduction index value of 38 dB? Assured that there was no chance of that, he blurted out that he had been given a test report as proof! The so-called test report was a simple sound reduction index curve without any trace of leakage. The acoustician insisted that a full test report should include the description of the sample, and the full report was eventually sent to the economist. On examining the full document, it turned out that the contractor had simply tested the door panel without its frame.

2.9.4.5 Roof

In the course of a large performance facility project, one of the roof manufacturers in competition boasted of better acoustic performances than competitors and had a laboratory test report to show it. Looking closely at the description it read, "The panels are laid on the mineral wool"; it turned out that many mechanical fixations had actually been omitted compared to the normal mounting procedure (and such a fixation would never have been accepted by the safety engineer).

2.9.4.6 Plasterboard Partition

A large round-robin test was held in the 1980s [58] on a basic plasterboard partition. Testing the supposedly same kind of plasterboard partition (i.e., made of materials from the same origin), the deviation was ±6 dB on the global value. The spread of values in the 125 Hz band could reach 25 dB.

2.10 BASICS OF REGULATIONS

2.10.1 Purpose

Regulations are meant to set a policy and specify limits. As opposed to standards, their application is compulsory. As an example, they may provide noise level limits for community noise control, transportation noise, or occupational noise control purposes. They may also provide minimal acoustic requirements for various types of buildings, such as dwellings and schools.

Due to the technicalities involved in Europe, the trend is more and more to rely on standards that are quoted and referred to in the legal texts.

One might care to remember that regulations are not meant for comfort! They simply provide the bare minimum acoustic requirements.

Regarding transportation noise, limits are stated using noise levels over specific spans of time (e.g., daytime or nighttime period). Equipment noise may be covered using noise level limits (e.g., [62]) or the emergence (gap between the ambient noise measured with the equipment in operation, and the background noise level without the equipment operating) (e.g., [61]).

2.10.2 Process

At the beginning there usually will be a law written by the national assembly (e.g., Noise Abatement Act in the United States [59] or Loi sur le bruit in France [60]). It will set the policy regarding certain topics (e.g., community noise control or building acoustics requirements). Next, there will be a governmental decree precisely stating the acoustic requirements (e.g., [61] in France). In addition, there will be a governmental (or even provincial or local) edict usually developing the guidelines and stating the means of verification of compliance as well as the possible fines should infringements be found. Last, there will be a governmental or provincial or local circular letter (e.g., [62]) explicating the contents of the edict, as it may vary from one place to another. U.S. examples are given in [63]. In some cases one may even have a global regulation (e.g., [64]) that is completed by ordinances stating noise limits (e.g., [65]) pertaining to a specific building or facility (e.g., [66]) or location (e.g., [67]).

2.10.3 A Few Difficulties

The reader has probably understood by now: One may very well boast of making a law, but it will not be of any influence as long as the decree has not been signed.

On the opposite, some local authorities may in certain cases lower the limit values by an edict.

Last but not least, while it may be desirable to have a single regulation on community noise, there may be some discrepancies according to the local customs and habits. For example, in Europe there eventually was an agreement on using a L_{den} noise level limit, but the notion of evening differs a great deal between, e.g., a Dutch person and a Spanish person, and the definition of those time spans can differ a great deal!

2.10.4 Using the Standards in the Regulations

Regulations can be quite complicated to change, and any change does take time. In Europe it is now quite usual to quote European standards (or even national standards) in a decree in order to simplify its writing (keeping in mind that it is easier to change a standard than a law text).

2.10.5 Expert to the Court

The expert is called by the court of justice to assist the judge in a technical matter. He will take stock of the situation and make his assessment. This may mean that he will point out not only which of the existing law texts are applicable, but also what extra rules he wishes to see applied.

2.10.6 Examples

2.10.6.1 Noise from Mechanical Equipment in a Building

The proud owners of a new French condominium quickly found that during the night some mechanical piece of equipment inside the condominium was quietly emitting a hissing sound. An expert was called, and it was found that the sound level values did comply with the applicable regulations for dwelling acoustics (i.e., 30 dB(A) in France); more to the point, those values did not even reach the minimal value required by the local authorities to start investigating a community noise complaint (i.e., 25 dB(A) in this region). Nevertheless, the court found the builders guilty, as the expert thought that this noise could be considered annoying when considered as community noise (with the offense being characterized as the noise level being greater than the background noise level by at least 3 dB(A) in France).

2.10.6.2 Noise from a Cinema

During a rehabilitation project an old Parisian cinema was found to generate noise in the newly built neighboring spaces. The expert who was assigned the case started by stating that though the regulations on musical venues have explicitly excluded the cinemas from their range, he would nevertheless use them in this case! This led to more stringent precautions than usual.

2.10.6.3 Noise from a Cock

Recently retired people bought a country house and quickly started to complain about the noise generated by the animals in the field next to their property. The cock was especially targeted as a nuisance source.

   The case was eventually brought to the court, which followed the expert's advice: In town you have to cope with the noise from motor vehicles, in the countryside you have to bear with the noise from animals.

2.10.6.4 What Is Activity Noise?

A comic scene was reported in a high-speed train when a businessman shouted at length in his cell phone without regard for the other passengers. Eventually a gentleman started humming and singing an opera air, louder and louder, until the exasperated businessman ordered him to clam up: "Can't you stupid bugger see I'm working?" he shouted. Back came the quiet answer: "Actually so am I sir, I am under contract with the opera." The applause of the whole carriage followed, as did the departure of the businessman.

Lesson Learned: What is normal activity noise for one party may be construed as sheer annoyance by others.

2.10.6.5 Rehabilitation of a Dwelling

An old Parisian flat was bought by new residents who decided on thoroughly improving it. In the lounge and dining room the old worn-out carpet on an old wooden floor was duly replaced by marble tiles on a floating screed, while the kitchen, complete with a dishwasher, was relocated closer to the dining room.

   The tenants of the flat underneath were quick to remark that they frequently had to cope with noise annoyance. An expert was mandated by the court. It was found that all the acoustic performances of the rehabilitated flat did reach the acoustic targets that would have been required for new construction. However, the expert pointed out that the relocation of the kitchen made for higher transmitted sound levels than before. He also pointed out that the replacement of the old carpet by a floating screed made for higher-impact sound levels than before. The decision of the court was to order the rehabilitated flat to be put back to its original state.

Lesson Learned: Aside from the usual regulations, one does have to take into account jurisprudence.

2.11 LABORATORY AND *IN SITU* MEASUREMENTS

2.11.1 Purpose

Acoustic measurements are needed to characterize the acoustic properties of materials, assemblies, or equipment. According to the required accuracy, various methods are available.

One thing must be clear from the beginning: A measurement result is only valid for a given setup (e.g., for certain dimensions and construction scheme of a material, even sometimes for specific time spans). A measurement result without any proper description (e.g., type, dimensions, environment of the sample) is simply useless! A so-called measurement report made of a simple numerical value or graphic must not be accepted. The relevant acoustic measurement standards provide a measurement procedure as well as a list of things to be reported.

2.11.2 Mastering the Measurement Conditions

Measurements can be affected by many factors, ranging from the type of measuring equipment to the laboratory or test site characteristics to the mounting conditions of the specimen under test, not to mention the operator's skills!

In order to master the measuring conditions, one usually needs an acoustic laboratory. There are standards defining the requirements for a laboratory for sound reduction index and impact sound purposes [68, 69], as well as in ISO 354 for sound absorption determination [73] and in ISO 3740 series for sound power determination [75].

In addition, the specimen under test must be mounted according to the standard's requirements. In the olden days, this was often described in the relevant measurement standard; nowadays, there is a tendency to try to define the mounting conditions in a separate standard usually titled "test code for X."

2.11.3 Making Do with Existing Conditions

There may be a couple of reasons to try to perform *in situ* measurements. The most obvious are:

- Commissioning measurements
- Testing a specific configuration or equipment hard to reproduce in the lab

In situ measurements are trickier than laboratory measurements: To start with, one does not necessarily enjoy the low background noise levels of a laboratory. Next, one will probably not enjoy as good of a diffuse field as in the lab. Last, flanking transmissions may be a problem for some of the measurements. Nevertheless, one might still get some good indications, keeping in mind a couple of basic rules:

- The measured signal should be 10 dB over the background noise level. One may even get as low as 6 dB over the background noise, but corrections are then needed on the result.
- The mounting conditions and the configuration of the premises must be properly recorded in order to enable the proper use of the measurement results.

2.11.4 Standards for Measurements

Measurement standards will typically describe the measurement procedure and the applicable standards for the measurement apparatus. They will also give guidelines regarding the evaluation of the uncertainty of measurement and list the various things to be reported, including the proper format of graphs and results to enable comparison between tests. Below are a few types of measurements.

2.11.4.1 Sound Reduction of a Wall or Floor

The sound reduction index measurement of a wall or floor will typically be performed under laboratory conditions. The test specimen is mounted in a test frame between two test rooms and submitted to the diffuse sound field in the emission room. The sound pressure levels are measured in the emission room and in the receiving room according to the ISO 140 series [70]. Measurements are also possible using sound intensity measurements on the receiving side according to ISO 15186 [71].

2.11.4.2 Impact Sound Level of a Floor

The impact sound level of a floor can be measured under laboratory conditions using an impact machine. The test specimen is mounted horizontally in a test frame between two reverberant rooms. Sound pressure levels are measured in the receiving room according to ISO 10140-3 [72].

2.11.4.3 Absorption Coefficient

The absorption coefficient can be measured under diffuse field conditions using a reverberant room according to ISO 354 [73]. It can also be measured under normal incidence using a wave tube according to the ISO 10534 series [74]. Due to the standardized procedure in which diffraction may appear on the edges of the sample and the sound field in the test room may be affected by a significant absorptive area some test reports will give a greater than 1 value.

2.11.4.4 Sound Power Level

Sound power level assessment can be performed using laboratory measurements, or even *in situ* measurements for engineering class measurements (taking into account the fact that some pieces of equipment are way too big to enter a laboratory!). Those measurements are performed using sound pressure level measurements either in a reverberant room or on an enveloping surface around the sound source of interest using the ISO 3740 series [75], or using sound intensity measurements according to the ISO 9614 series [76, 77].

2.11.4.5 Measurement and Computation

Why bother to measure when one can compute? To start with, there will probably be a bit of a difference between the predicted result using simple formulas whose validity domain usually calls for homogeneous materials, and the actual specimen, due to (nonexhaustive list) nonhomogeneities of the materials, imperfect mounting conditions, and even noncompliance of the product with its commercial description. While a computation probably will help to restrict the choices that had been left opened (e.g., regarding mounting conditions and choice of materials) by comparing specimens, ultimately a measurement will be needed to ascertain the actual absolute value (even taking into account some limitations induced by the laboratory conditions). More to the point, one will be able to show the test to the client, bearing in mind that as the saying goes, "Nobody believes a calculation except the person who did it, and everybody believes an experiment or measurement except the person who did it."

Incidentally, this is as good a time as ever to make sure that the contractor will be able to perform the required task and the manufacturer will be able to provide the required technical assistance. For difficult projects, there probably will be a laboratory test to physically validate the solution, and later a proof room construction to enable the architect and the end user to see the result. (Note: When properly built, it will also enable the acoustician to perform measurements on some elements that may have been subjected to limited laboratory tests, e.g., the façade, due to larger dimensions than those acceptable in the laboratory.) Ultimately, a head of series will be constructed in the actual building under construction; measurements will help validate its implementation by the contractor under real conditions (better discover at this stage that the operating procedure is wrong than after a hundred such constructions have been completed!).

Note: It is extremely tempting for the end user and the contractor to save time and money on such niceties as airproofing. Should that be the case, a valuable test opportunity will be lost.

2.11.5 Examples

2.11.5.1 Variability with Time

In the end of the 1980s, a round-robin test was undertaken in France regarding the measurement and prediction of the spatial sound level decay in a room [15]. Due to the high number of participating teams, the measurements were spread from June to November with the institute opening the process. On completion it was found that the results of the last months differed from the earlier ones. Being suspicious of the accuracy of the participating technicians involved, the institute performed a new set of measurements; it discovered that its results also differed! This was eventually traced to the hygrometry and temperature.

Lesson Learned: Always write down the measurement conditions.

2.11.5.2 Traffic Noise

An engineering outfit had to assess the noise levels generated by a large transportation corridor near its pet project. Due to the urgency, an occasional technician (i.e., an office helper who had been shown the basic sound level meter measurement procedure) was instructed to perform the measurements. On looking at the results, the engineers were puzzled: The sound level values were within 2 dB(A) of each other. Interrogating the technician, they eventually found that the measurements had been performed from the car due to the low temperature. One engineer then went for the kill: Had the engine been stopped during measurements? A dumbfounded technician then answered that one engine among so many others would not change much, would it?

Lesson Learned: Always explain the reasons for performing the measurements a given way; try to have a rough idea of the range in which the measurement results should be found.

2.11.5.3 Reverberation Time Measurements

Reverberation time measurements can be carried out using a powerful sound source that will excite the various modes of a room prior to cutting it out and assessing the time decay. However, for practical purposes, smaller portable impulsive sound sources have often been used. The explosion of balloons can give results in this matter. However, in the olden days it was common to use blank cartridges in a firearm. This led to several funny or nasty incidents. The author personally experienced a few of them: on measuring the reverberation time in a school restaurant, a paper bird came slamming down on the nearest table. In another episode, in a fashionable restaurant the old main lighting fixture that was awkwardly suspended to the ceiling came crashing down after the shot. Slightly less funny (at least for maintenance), in an office tower restaurant the blank shot triggered the fire alarm that had not been set to safe, and all fresh air trapdoors on the façade automatically opened (and had to be manually closed). Nowadays, security measures have put an end to such a way of measuring.

2.11.5.4 Door Panel

A puzzled engineer once called an acoustician friend of his to ask whether a door with a gap underneath could achieve a sound reduction index of 30 dB. On being answered negatively, he then announced that he had been given a test report by the contractor boasting of 35 dB. That "test report" happened to be a mere sound reduction index versus frequency curve without any description of the measurement conditions; the fact that there was no trace of leakage (which would have induced a flattening of the curve) was highly suspicious too. After much shouting on the phone, the full measurement report was eventually handed over. It turned out that the test had been performed on the door panel alone (without any frame or gap for that matter).

Lesson Learned: Do not trust an incomplete test report.

2.11.5.5 Roof Assembly

A manufacturer had in his catalog a roofing system that apparently featured better acoustic performances than the competition.

On looking earnestly at the test report, one sentence read: "The panel was laid on the structure." This was the clincher: There was no fastener, which helped gain a few dB, and would of course never work in practice.

Lesson Learned: Always read carefully the test sample description as well!

2.11.5.6 Ceiling

A manufacturer's representative would boast that his ceiling was simultaneously highly absorptive and featured high insulation performances. When requiring the relevant test reports, a technical file was sent back. The relevant test reports were printed out in small characters (usually a bad sign for sure!), the test laboratories were different for the absorption test and for the insulation test, and the name of the product appeared to have been tampered with. It eventually turned out that the representative had taken his best absorptive ceiling and his best insulating ceiling and attempted to pass them for a single material.

Lesson Learned: Reading a test report often is a case of finding where the clincher is!

2.12 ACTIVE NOISE CONTROL

2.12.1 Active Noise Control

This section does not intend to cover what now is a broad subject. On the other hand, due to the numerous inquiries about the mysterious "antinoise" that happens in a day's work, some explanations are in order.

On paper it is simple and straightforward: A noise is made of numerous sinusoidal waves; in order to cancel that noise, one simply has to generate the required opposite phase sinusoidal waves. The idea is so simple that it was already outlined in papers by the Persian physicist Al Farhabi in the 14th century. Later on, at the end of the 19th century, Helmholtz put the problem into equations. The actual experiment had to wait until the war preparations of 1933: At that time the detection of planes was mainly performed by acoustic means; thus, silencing them would give a clear advantage. Accordingly, the German physicist Olsen attempted active noise cancellation on a stationary source: a transformer. It quickly turned out that while a significant gain could be reached at a given point, amplification would be produced in other points too due to the impossibility of physically superposing the actual noise source and the counternoise source. With the development of radar, work on active noise control lost its immediate interest and was stopped for the time being.

Interest resumed again in the mid-1970s with its application on a single-dimension system, as there is no interference. In a ventilation duct the use of an active system helps avoid pressure loss and is economically viable when compared to a standard 3 m long dissipative silencer. Later on, the development of computational methods opened new perspectives in 3D.

The simplest active noise cancellation system calls for a back-feed loop where one checks that the overall resulting sound level value is minimized. The signal is picked when it arrives and phase reversed before being injected in a loudspeaker. In more elaborate systems, the signal is associated with several parameters (e.g., throttle setting), and a transfer function has been elaborated so as to produce the correct cancellation signal.

2.12.2 Examples

2.12.2.1 Headphones

A few manufacturers now have headphones with in-built active cancellation to enhance the sound quality of the speech or music signal for the listener [78].

2.12.2.2 Silencers on Air Intake

Air intakes are important features of a building, as they allow fresh air inside; however, they can constitute a serious acoustic weakness in the façade insulation. To minimize the potential complications of a long silencer, some manufacturers have introduced active cancellation inside the duct [79].

2.12.2.3 Airplanes

One has probably experienced active noise cancellation traveling on some airplanes (e.g., the ATR family or the B-777), where the sound levels inside the cabin are controlled using actuators on the walls.

REFERENCES

1. R.B. Lindsay, "Lindsay's Wheel of Acoustics," Created by R. Bruce Lindsay, *Journal of the Acoustical Society of America*, 36, 2242 (1964).
2. *Livre blanc de l'acoustique en France* (white book on acoustics in France) (in French), SFA, Paris, 2010, https://www.sfa.asso.fr/fr/documentation/livre-blanc-.
3. F.V. Hunt, *Origins in acoustics—The science of sound from antiquity to Newton*, Yale University Press, New Haven, CT, 1978.
4. C. Gabriel, Cours d'acoustique (in French), http://www.claudegabriel.be/Acoustique.html.
5. P. Liénard, *Petite histoire de l'acoustique—Bruits sons et musique* (*Small history of acoustics—Noises sounds and music*, in French), Hermès-Lavoisier, Paris, 2001.
6. M. Asselineau, *Quelques éléments d'histoire de l'acoustique* (*A few elements on the history of acoustics*, in French), Paris, 1984.
7. M. Mersenne, *L'Harmonie universelle*, 1636; reedited by CNRS, Paris, 1986.
8. IEC 537: *Frequency weighting for the measurement of aircraft noise (d-weighting)*, International Electrotechnical Commission (IEC), 1976.
9. IEC 61672: *Electroacoustics—Sound level meters*, International Electrotechnical Commission (IEC), 2002.
10. H. Fletcher, W.A. Munson, Loudness, its definition, measurement and calculation, *Journal of the Acoustical Society of America*, 5, 82–108 (1933).
11. B. Berglund et al., *Guidelines for community noise*, WHO, 1995.
12. ISO 1996:1971: *Acoustics—Assessment of noise with respect to community response*, Publication R 1996, Geneva, 1971.
13. ANSI S12-2-2008: *NC curves*.
14. M. Asselineau, Quelques aspects de l'affaiblissement acoustique d'un élément de façade soumis au bruit de la circulation (A few aspects of the sound reduction of a facade element exposed to traffic noise, in French), doctoral thesis, Le Mans, 1983.
15. A.M. Ondet, J.L. Barbry, *Prévision des niveaux sonores dans les locaux encombrés* (*Prediction of sound levels in fitted rooms*), NST0052, INRS, Nancy, 1984.
16. A.M. Ondet, J.L. Barbry, *Acoustique prévisionnelle—Modélisation de la propagation dans les locaux industriels encombrés à partir de la technique des rayons-logiciel RAYSCAT* (*Predictive acoustics—Modeling of sound propagation in fitted industrial rooms using ray tracing—RAYSCAT program*), NST0067, INRS, Nancy, 1987.
17. Odeon room acoustic software, http://www.odeon.dk/.
18. CATT (Computer Aided Theater Technique) software, http://www.catt.se/.
19. J.G. Migneron, *Acoustique urbaine*, Masson Paris, 1983.
20. W.M. Schuller et al., *Contrôle du bruit en milieu industriel* (*Industrial noise control*, in French), Dunod Paris, 1981.
21. Peutz, ILBOS program, Nijmegen, 1984.
22. I. Ver, L. Beranek, *Noise and vibration control engineering*, J. Wiley & Sons, New York, 1992.
23. ISO 9613-2: *Acoustics—Attenuation of sound during propagation outdoors—Part 2: General method of calculation*, Geneva, 1996.
24. M. Pierette, E. Parizet, P. Chevret, Perception and evaluation of noise sources in an open space, presented at Proceedings of ICA2013, Montreal.
25. ISO 17624: *Acoustics—Guidelines for noise control in offices and workrooms by means of acoustical screens*, Geneva, 2004.

26. S. Viollon, C. Lavandier, C. Drake, Influence of visual setting on sound ratings in an urban environment, *Applied Acoustics*, **63**(5), 493–511 (2002).

27. A. Gidlöf-Gunnarsson, E. Öhrström, Noise and well-being in urban residential environments: The potential role of perceived availability to nearby green areas, *Landscape and Urban Planning*, 83, 115–126 (2007).

28. S.H. Tang et al., Monte Carlo simulation of sound propagation through leafy foliage using experimentally obtained leaf resonance parameters, *Journal of the Acoustical Society of America*, 67, 66–72 (1980).

29. R.W. Guy, The transmission of sound through walls, windows, and panels: A one dimensional teaching model, *Canadian Acoustics*, 12(4), (1984).

30. TNO, LGILAB program, Peutz, 1984.

31. L. Beranek, On reverberation, presented at ICA 2013 Proceedings, Montreal, 2013.

32. EN/ISO 14257: *Acoustics—Measurement and parametric description of spatial sound distribution curves in workrooms for evaluation of their acoustical performance*, Geneva, 2001.

33. D.L. Steele, S.H. Chon, *A perceptual study of noise annoyance*, AudioMostly, 2007.

34. Projet de guide indicateurs physiques acoustiques et vibratoires adaptés au ressenti des riverains (*Project for a guide on acoustic and vibratory physical indicators suitable to the feelings of the neighbours*, in French), Afnor S30MI Final Report, AFNOR, Saint Denis, 2014.

35. M. Concha-Barrientos et al., *Occupational noise—Assessing the burden of disease from work related hearing impairment at national and local levels*, Environmental Burden of Disease Series 9, WHO, 2004.

36. Passchier-Vermeer, *Noise exposure and public health*, TNO, 1993.

37. R. Guski et al., The concept of noise annoyance: How international experts see it, *Journal of Sound and Vibration*, 223(4), 513–527 (1999).

38. F. van den Berg et al., The relation between scores on noise annoyance and noise disturbed sleep in a public health survey, *International Journal of Environmental Research and Public Health*, 11, 2314–2327 (2014).

39. D.C. DeGagne, A. Lewis, Criteria to minimize noise annoyance from industrial applications, http://www.noisesolutions.com/uploads/images/pages/resources/pdfs/Noise%20Annoyance%20Paper.pdf.

40. D.S. Michaud et al., Noise annoyance in Canada, *Noise and Health Journal*, 7(27), 39–45 (2005).

41. J. Lambert, C. Philipps-Bertin, Perception and attitudes to transportation noise in France: A national survey, presented at Proceedings of the 9th International Congress on Noise as a Public Health Problem, Foxwoods, 2008.

42. S. Viollon, Influence des informations visuelles sur la caractérisation de la qualité acoustique de l'environnement urbain (Influence of visual informations on the characterization of the acoustic quality of urban environment, in French), doctoral thesis, Cergy, 2000.

43. P. Schomer et al., *A white paper: Assessment of noise annoyance*, Schomer & Associates, 2001, http://www.nonoise.org/library/schomer/assessmentofnoiseannoyance.pdf.

44. D. Schreckenberg et al., Annoyance and disturbances due to traffic noise at different times of day, presented at Proceedings of Inter Noise 2004, Prague.

45. MRSC, Seattle, WA, http://www.mrsc.org/search/searchresults.aspx?q=noise%20regulations.

46. City of Belmont, Noise ordinance, 2004.

47. Elmbridge Borough Council, What is a statutory noise nuisance? http://www.elmbridge.gov.uk/envhealth/noise/statnuisance.htm#sthash.a6Ml1zux.dpuf.

48. T.J. Schultz, Synthesis of social surveys on noise annoyance, *Journal of the Acoustical Society of America*, 64(2), 1978.

49. M. Asselineau, Quelques aspects de l'affaiblissement acoustique d'un élément de façade soumis au bruit de la circulation (A few aspects of the sound reduction index of a facade sample exposed to traffic noise, in French), doctoral thesis, Le Mans, 1987.

50. *Survey of legislation, regulations, and guidelines for control of community noise—Final report of the I-INCE technical study group on noise policies and regulations*, Publication 09-1, International Institute of Noise Control Engineering, 2009.

51. Traffic noise reduction in Europe, 2008, www.transportenvironment.org/sites/te/files/media/2008–02_traffic_noise_ce_delft_report.pdf.
52. A. Gunnarsson, M. Ögren, T. Jerson, E. Öhrström, Railway noise annoyance and the importance of the number of trains, ground vibration and building situational factors, *Noise and Health Journal*, 14(59), 190–201 (2012).
53. C. Eriksson, M.E. Nilsson, G. Pershagen, *Environmental noise and health—Current knowledge and research needs*, Report 6553, Naturvardsverket, Stockholm, 2013.
54. K. Holmberg, Critical noise factors and their relation to annoyance in working environment, Doctoral thesis, Arbetsvetenskap/Miljöteknik, LuLea Universitet, 1999.
55. FDIS ISO 1996: *Acoustics—Description, measurement and assessment of noise in the environment*, Geneva, 2014.
56. Recommandation du Ministère de la Santé du 21 juin 1963 relative au bruit (Recommendation of the French Ministry of Health on noise, in French), Paris, June 1963.
57. AFNOR, General guidelines, Saint Denis, 2009.
58. R. Pompoli, *Report on round robin test of transmission loss of plasterboard partitions*, Bologna, 1980.
59. Noise Control Act of 1972, PL 92-574, 86 Stat. 1234, 42 USC §4901.
60. Loi n° 92-1444 du 31 décembre 1992 relative à la lutte contre le bruit (Law on noise control, in French), *Journal Officiel de la République Française*, January 1, 1993.
61. Décret **n°** 2006-1099 du 31 août 2006 relatif à la lutte contre les bruits de voisinage et modifiant le code de la santé publique (dispositions réglementaires) (Decree pertaining to community noise control, in French), *Journal Officiel de la République Française*, September 1, 2006.
62. *A guide to New York City's noise code*, NYC Environmental Protection, New York, 2012.
63. R.C. Chanaud, Noise ordinances—Tools for enactment modification and enforcement of a community noise ordinance, www.noisefree.org/Noise-Ordinance-Manual.pdf.
64. Réglement sur le bruit de la Ville de Montréal (Noise regulations of the City of Montreal, in French), SVPM, Montréal, 1994.
65. Ordonnance n° OCA09 001—Arrondissement de Ahuntsic-Cartierville (Ordinance Ahuntsic-Cartierville area, in French), RRVM, chapitre B-3, article 20, Montréal, 2009.
66. Ordonnance n° OCA13 005—Marché de quartier Station de métro Sauvé (Ordinance on market, in French), RRVM, chapitre B-3, article 20, Montréal, 2013.
67. Ordonnance n° OCA10-(B3)-001—Arrondissement de Rivière des Prairies Pointe aux Trembles (Ordinance on area Rivière des Prairies, in French), RRVM, chapitre B-3, article 20, Montréal, 2010.
68. ISO 140-1: *Acoustics—Measurement of sound insulation in buildings and of building elements—Part 1: Requirements for laboratory test facilities with suppressed flanking transmission*, Geneva, 1997.
69. ISO 140-2: *Acoustics—Measurement of sound insulation in buildings and of building elements—Part 2: Determination, verification and application of precision data*, Geneva, 1991.
70. ISO 14003: *Acoustics—Measurement of sound insulation in buildings and of building elements—Part 3: Laboratory measurements of airborne sound insulation of building elements*, Geneva, 1995.
71. ISO 1518603: *Acoustics—Measurement of sound insulation in buildings and of building elements using sound intensity—Part 3: Laboratory measurements at low frequencies*, Geneva, 2002.
72. ISO 1014003: *Acoustics—Laboratory measurement of sound insulation of building elements—Measurement of impact sound insulation*, Geneva, 2010.
73. ISO 354: *Acoustics—Measurement of sound absorption in a reverberation room*, Geneva, 2003.
74. ISO 10534-1: *Acoustics—Determination of sound absorption coefficient and impedance in impedance tubes—Part 1: Method using standing wave ratio*, Geneva, 1996.
75. ISO 3740: *Acoustics—Determination of sound power levels of noise sources—Guidelines for the use of basic standards*, Geneva, 2000.
76. ISO 9614-1: *Acoustics—Determination of sound power levels of noise sources using sound intensity—Part 1: Measurement at discrete points*, Geneva, 1993.
77. ISO 9614-3: *Acoustics—Determination of sound power levels of noise sources using sound intensity—Part 3: Precision method for measurement by scanning*, Geneva, 2002.
78. Sennheiser, Commercial brochure, 2011.

79. Aldès, Entrées d'air et silencieux actifs (Air intakes and active silencers, in French), Commercial brochure, 2003.

Chapter 3

Building Acoustics

3.1 FOREWORD

When an incident sound wave arrives onto a material, part of it can be:

- Reflected
- Transmitted through the material
- Absorbed inside the material

This leads to the definition of:

- A reflection coefficient ρ defined as the ratio of reflected energy over incident energy
- A transmission coefficient τ defined as the ratio of transmitted energy over incident energy
- An absorption coefficient α defined as the ratio of reflected energy over incident energy

$$\rho^2 + \tau^2 + \alpha^2 = 1$$

One may care to note that those coefficients are frequency dependent. Usually the higher the frequency, the lower the transmission coefficient. More to the point, usually with soft or porous materials, the higher the frequency, the higher the absorption coefficient.

3.2 INTRODUCTION

As implied by its name, the field of building acoustics covers all aspects involved in the construction of a building. Usually such acoustics mainly revolve around a same concern: enabling the occupants to be decently protected from the noise and vibration aggressions from the outside environment, ensuring proper sound insulation between the various spaces of the building, controlling the reverberation inside the premises, and controlling the noise from mechanical equipment. In addition, one will also be concerned with the noise radiated to the environment either from the mechanical equipment or from the activities carried out inside the building.

This chapter is devoted to the basic physics of those phenomena. Application of those bases will be done in the relevant following chapters.

3.3 SOUND INSULATION

3.3.1 Experiencing Sound Insulation

Whoever has been living in flats certainly has firsthand experience in the notion of sound insulation when woken up in the middle of the night by noise from the next door flat! One may even have observed that according to the nature of the walls and the dimensions of the room, the perceived sound insulation was different.

3.3.2 Sound Reduction Index

The sound reduction index of a wall or floor is measured in a laboratory that features heavy slabs and walls with a framed 10 m² opening in the middle. Due to the constitution of the envelope of the measuring rooms that are separated by an expansion joint, flanking transmissions can be considered negligible, and the sole contribution to the sound levels measured in the receiving room comes from the radiation of the floor or wall under test. The floor or wall under test is excited using a diffuse sound field that is generated using a couple of loudspeakers located in one room (labeled "emission room") and directed toward the corners of the room opposite of the wall under test; the mean sound pressure level is measured in both the emission room and the receiving room.

The sound reduction index R is given by

$$R = L_1 - L_2 + 10 \log (S/A)$$

where L_1 is the mean sound pressure level in the emission room, expressed in dB, L_2 is the mean sound pressure level in the reception room, expressed in dB, S is the area of the floor or wall specimen under test, expressed in m², and A is the equivalent absorptive area in the reception room, expressed in m².

Measurements are performed either in the 100–3150 Hz range (ISO) [1] or in the 125–4000 Hz range (ASTM) [2]. More recently, EN/ISO standards have attempted to extend the range in the low-frequency region [3] down to the 50 Hz third octave band.

Of course, the measurement of the sound reduction index of a building component must be performed under laboratory conditions. But it may be possible to perform such a measurement *in situ* using sound intensity measurements as long as some conditions are met (mainly a low and stable background noise, and either a nonreverberant environment on the receiving side or reflective walls at least 3 m from the element under investigation) [4].

Note: When performing a sound reduction index measurement on a wall, roof, or floor sample, bear in mind that the dimensions of the sample should be as close as possible to the dimensions used in the project. Using smaller dimensions will usually result in overestimating the performance in the low-frequency range.

3.3.3 Sound Reduction Improvement Index

The addition of a doubling on an existing wall or floor will, of course, have consequences on the value of the sound reduction index. The effect of a doubling is characterized by the sound reduction improvement index ΔR. The value of ΔR is obtained through the measurement of the sound reduction index of the main wall, and the measurement of the sound reduction index of that wall with its doubling:

$$\Delta R = R_{wall\ with\ doubling} - R_{wall\ alone}$$

Of course, the lighter the wall, the larger the value of ΔR will be. In addition, the more porous the wall, the larger the value of ΔR will be. The usual walls used for a test are a 16 cm plain concrete wall, a 20 cm brick wall, and a 20 cm hollow concrete block wall. Plaster blocks 7 cm thick may also be found (e.g., [82]).

Please do note that the sound reduction improvement index ΔR can feature a negative value (i.e., –4 to –7 dB) for a plasterboard on polystyrene doubling glued on a brick wall. Whenever using a doubling, one must make sure that it will not hamper the overall sound reduction performance of the wall construction under consideration.

3.3.4 Sound Insulation

The sound insulation between two rooms will depend not only on the direct transmission through the separating wall or floor, but also on the flanking transmissions by all walls or floors linked to this separating element. Last, parasite transmissions (such as leakage around a constructive element or through a duct or opening) will also play a role. Figure 3.1 displays a few transmission paths between two rooms.

The sound insulation between rooms usually is measured using a diffuse sound field that is generated using a couple of loudspeakers located in one room (labeled "emission room"); the mean sound pressure level is measured in both the emission room and the receiving room.

The sound level difference D is given by

$$D = L_1 - L_2$$

where L_1 is the mean sound pressure level in the emission room, expressed in dB, and L_2 is the mean sound pressure level in the reception room, expressed in dB.

However, this sound insulation value is sensitive to the amount of acoustic absorption inside the receiving room: This would eventually mean that according to the fittings in the room, one would pass or fail the sound insulation criteria. In order to avoid this inconvenience, the usual rule is to standardize the result by the reverberation time.

The standardized sound insulation D_{nT} is given by

$$D_{nT} = L_1 - L_2 + 10 \log (T/T_0)$$

where L_1 is the mean sound pressure level in the emission room, expressed in dB, L_2 is the mean sound pressure level in the reception room, expressed in dB, T is the reverberation

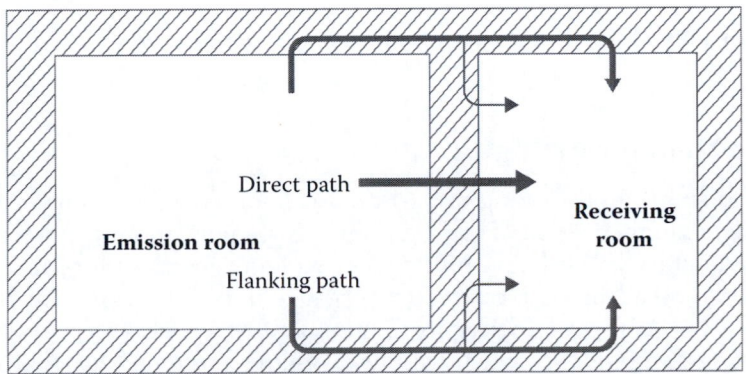

Figure 3.1 Direct and flanking transmission between rooms.

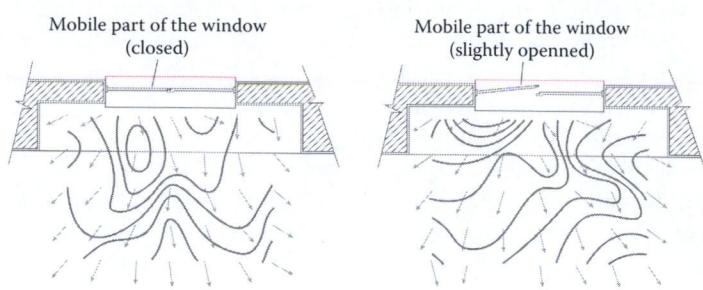

Figure 3.2 Example of sound intensity contours behind a window, either closed (left) or slightly opened (right), according to [4].

time in the receiving room, expressed in s, and T_0 is the reference reverberation time (usually 0.5 s).

For many years the usual frequency measurement range used to be 100–3150 Hz (ISO) [5] or 125–4000 Hz (ASTM) [6], with some other countries using the 100–5000 Hz (e.g., France [7]). More recently, EN/ISO standards have attempted to extend the range in the low-frequency region [8] to the 50 Hz third octave band.

Note: The choice of frequency range is not a purely academic matter; limiting the higher-frequency value will often result in avoiding the presence of a badly located coincidence frequency in the expression of the result, and influence the final single number rating (cf. Section 3.5).

While inside a building space the preferred method will be to use a loudspeaker turned toward a corner to help generate as close as possible a diffuse field; when attempting to measure the sound insulation of the façade things get more complicated: One is no longer in diffuse field conditions, and the choice is between using a loudspeaker sound source or the actual traffic noise [9]. Due to the coincidence effects, one has to carefully locate the sound source according to the standard recommendations (incidentally, make sure both the neighbors and the police are forewarned of the measurement too!).

Note: A construction project very often features a proof room that will show to the architect and the end user how the room will look. When required, in time it can be used for acoustic testing purposes too, but kindly remember that any leakage in the envelope of this proof room will result in unsatisfactory acoustic performance, and the contractor must be made aware of it.

Figure 3.2 displays the effect on the sound field of a room when a window is slightly opened or leaky.

3.3.5 Single Number Rating

In order to help speed up the rating of partition or floor assemblies, one usually relies on a so-called single number rating. This is an obvious simplification, as instead of working on the usual six octave bands (125–4000 Hz) of building acoustics, one now uses a single value. Now for the bad news: Nearly every organization has its own system. To try to simplify things, the American Society for Testing and Materials (ASTM) uses the Sound Transmission Class (STC) [10], which features a sliding contour. This contour runs 9 dB per octave from 125 to 500 Hz, 3 dB per octave from 500 to 1250 Hz, and stays flat from 1250 to 4000 Hz. One glides the contour over the sound reduction or sound insulation curve tested until the sum of negative deviations is no greater than 32 dB and the maximum negative deviation is 8 dB. ISO uses the R_w or D_w system [11], which is nearly identical, save that the starting point is at 100 Hz and the endpoint at 3150 Hz. In order to cope with European regulations, there usually are two values between parentheses next to it that are respectively given the symbols C and C_{tr} in order to obtain the value expressed for a pink noise and a traffic noise, respectively. So, when the sound reduction index value of a window is announced using a single number, one had better make sure of the nature of the indicator used.

The reader has understood by now that according to the spectrum emitted the sound source used for the tests, the single number rating value will not be the same. Some European countries (e.g., France and Spain) favor a traffic noise for the expression of façade sound insulation results, and the numerical value typically is 3 to 7 dB lower than the corresponding R_w value (see Section 3.12.6).

In order to cope with low frequencies down to 50 Hz, ISO has extended its R_w contour, with the relevant definition in standard ISO 16717-1 [12].

Note: Do be careful when dealing with a single number rating, as its value will depend on its definition (cf. Section 3.12.6).

3.3.6 Computing the Sound Reduction Index and the Sound Insulation

3.3.6.1 Sound Reduction Index

The crudest model is known as mass law. The sound reduction index R for a single wall is given by

$$R = 20 \log(mf) - 48$$

where m is the mass per area unit of the wall or floor, expressed in kg/m², and f is the frequency, expressed in Hz.

There are a few computer programs available to try to compute the sound reduction index of a wall or a floor [13, 14]. Usually such programs rely on the mass law but also quite often on a reasonable database to help tune the relevant parameters.

When dealing with multiple layers (e.g., a massive wall with a glued doubling made of a mineral wool factory assembled on plasterboard) or with multiple walls, there will be a resonance appearing for the gap between the plates. From this resonance frequency, the acoustic performance is much better (e.g., 12 dB per octave for a double wall) until dips from coincidence effects appear. More to the point, the type of connection between leaves of the wall will play a significant role (see Section 3.12.8). Usually, a multiple wall will fare much better than a single wall in the medium- and high-frequency range, but it can also come out the loser in the lower-frequency range.

A word of advice there: Whenever one tries to assess the sound reduction index of a wall or floor, first try to compute the sound reduction index of a known construction that was subjected to a laboratory test. Only then will you be able to try to start changing such parameters as mass and stiffness. Change one parameter at a time to check its influence. Do not hesitate to wonder and ask questions (e.g., it is not unusual on some computation models to observe a significant sound reduction index value increase when adding mineral wool on one side of a single wall; you know, of course, that it is ridiculous and does not happen in the real world!).

3.3.6.2 *Sound Insulation*

There are a few computer programs available to try to compute the sound insulation between rooms [15]. Such programs rely on the knowledge of the sound reduction index of the various walls, as well as their coupling parameters, and operate according to ISO 12354 [16]. Most of the time they come with a rather significant database.

A word of advice there: Should one of the constructive elements be missing from the database, whenever possible proceed as with the previous chapter, using as a basis as close as possible a material to try to get plausible coupling factors.

Looking at the results, one may try to modify one parameter at a time (e.g., increase the performance of the separating wall or floor) in order to check its influence. Usually the results can be displayed graphically so one may pinpoint which component needs upgrading.

While such computer programs are now quite user-friendly and reliable when it comes to concrete construction and dimensions similar to those encountered in dwellings, they can get trickier and more prone to errors when it comes to light construction or wood construction, or larger dimensions. There are currently developments in this field to try to cover wood constructions that are quite fashionable nowadays [17].

3.3.7 Noise Radiated by a Construction

The noise radiated by a building will of course depend on the sound reduction index of each envelope component as well as its respective area.

The sound power level L_w radiated by a component of area S and sound reduction index R is given by [18]

$$L_w = L_p - 3 - R + 10 \log(S)$$

where L_p is the sound pressure level inside the building.

There are a few things to get from this formula: To start with, large elements with a rather small sound reduction index value (e.g., the metal roof of a building) will be major contributors. So will the apertures (that can be considered an element with $R = 0$) and weaknesses with regards to other components.

3.3.8 Standards and Regulations

For a long time, there were various ways of expressing the acoustic requirements for buildings, especially dwellings, and a rather broad range of values to go with it [19, 20]. Nowadays, there are quite a number of standards and regulations pertaining to sound exposure levels and sound insulation [21, 22]. A few of them are given in the various chapters of this book regarding the main types of constructions. Typically, one may care to note the following constraints:

- There are limits, expressed as noise levels, regarding the noise exposure of workers (e.g., [23]), or the danger limit for people exposed to high-level music (e.g., [24]).
- There often are minimum sound insulation values required for specific types of construction, such as dwellings, schools, hospitals, etc. (e.g., [20, 21, 24–39]).

3.4 IMPACT NOISE

3.4.1 Experiencing Impact Noise and Walking Noise

Everyone probably has experience in the notion of impact sound transmission when woken up by the heavy steps of the occupant upstairs or simply by the steps of someone in another room within the same flat. One may even have observed that according to the nature of the floor, the walls, and the dimensions of the room, the perceived impact noise was different.

One has probably also observed that according to the nature of the floor covering or floor constitution, the noise of the steps inside the room where one is walking may sound quite differently (e.g., a wooden floor on joists will sound more hollow than a plain concrete floor).

3.4.2 Impact Noise of a Floor

The impact noise of a floor is measured in a laboratory that features a heavy slab with a framed 10 m² opening in the middle. Due to the constitution of the walls, flanking transmissions can be considered negligible, and the sole contribution to the sound levels measured in the receiving room comes from the radiation of the test floor. The floor under test is excited using an impact machine (also designated the tapping machine).

The impact noise L_n is given by

$$L_n = L_1 + 10 \log (S/A)$$

where L_1 is the mean sound pressure level in the reception room, expressed in dB, S is the area of the floor specimen under test, expressed in m², and A is the equivalent absorptive area in the reception room, expressed in m².

The standard describing the measurement procedure is usually ISO 140-6 [40] or ASTM E492 [41]. Those standards typically consider either the 100–3150 Hz range (ISO) or the 125–4000 Hz range (ASTM). More recently, EN/ISO standards have attempted to extend the range in the low-frequency region [42] down to the 50 Hz third octave band.

3.4.3 Impact Noise Transmission

The impact noise transmitted into a room will depend not only on the direct transmission through the separating floor, but also on the flanking transmissions by all walls linked to this separating element. Last parasite transmissions (such as leakage around a constructive element or through a duct or opening) will also pollute the result.

The impact noise transmitted into a room usually is measured using an impact machine on the floor of one room (labeled "emission room"); the mean sound pressure level is measured in the receiving room. The relevant measurement standard usually is ISO 140-7 [43] or ASTM 1007 [44].

The impact noise level is directly measured in the receiving room. However, this sound level value is sensitive to the amount of acoustic absorption inside the receiving room: This would eventually mean that according to the fittings in the room, one would pass or fail the impact noise criteria. In order to avoid this inconvenience, the usual rule is to standardize the result by the reverberation time.

The standardized impact noise level L_{nT} is given by

$$L_{nT} = L_1 + 10 \log (T/T_0)$$

where L_1 is the mean sound pressure level in the reception room, expressed in dB, T is the reverberation time in the receiving room, expressed in s, and T_0 is the reference reverberation time (usually 0.5 s).

The presence of a ceiling underneath the floor will of course affect the results. Some standards do take this into account (e.g., ASTM E1007 [44]).

The good news is the tapping machine is reproductive enough. More to the point, it does manage to provide useful data for impact noise prediction, for example, using standard ISO 15712-2 [45]. Now for the bad news: While it is a useful tool for the acoustician, it has trouble correlating with the eventual annoyance from impact sound. This is due to the fact that the generated impact spectrum is poor in the low-frequency range (which is not that surprising, as it was initially developed to try to imitate the impacts from high-heeled shoes). Unfortunately, one may have experienced that a bare-footed person walking on tile does not generate any greater impact noise than walking on a carpet, as this mainly is a low-frequency problem; yet the tapping machine will make a sizable difference out of it. In order to introduce this kind of impact, some countries have been using a rubber ball or even a small tire [46] as an impact sound source. Such a procedure has been investigated [47, 48] and is now considered in an EN/ISO project [49].

3.4.4 Rain Impact Noise

Rain impact noise can be quite significant on light roof elements. In order to help users compare the acoustic performances of roof assemblies or even constructive elements (e.g., roof trapdoors), a test procedure in the laboratory has been developed in standard ISO 140-18 [50].

Of course, the softer the roof covering, the quieter the rain impact noise will be. For modern designs it may be possible to use vegetated roofs [51] that provide a higher sound reduction index value and better impact performances than the standard roofs, while featuring extra acoustic absorption and sustainable development qualities.

3.4.5 Walking Noise of a Floor Assembly

Impact noise transmission concerns the case of the impact machine located on the floor of one room with the sound level measurement held in the receiving room. The walking noise of a floor assembly is assessed by performing the measurement in the same room. Of course, one has to make sure that the noise generated by the impact machine is smaller by at least 10 dB than the walking noise radiated by the floor (this may actually entail some adaptations to the impact machine).

3.4.6 Single Number Rating

In order to help speed up the rating of floor assemblies, one usually relies on a so-called single number rating. This is an obvious simplification, as instead of working on the usual six octave bands (125 to 4000 Hz) of building acoustics, one now uses one single value. Now for the bad news: There are a few rating systems around. To try to simplify things, ASTM uses the Impact Insulation Class (IIC), which features a sliding contour [52]. The curve of interest (sound reduction or sound insulation) runs flat until 250 Hz, and then descends 3 dB per octave to 1000 Hz, and 13 dB per octave from 1000 Hz to 4000 Hz. One glides the contour over the sound reduction or sound insulation curve tested until the sum of positive deviations is no greater than 32 dB and the maximum positive deviation is 8 dB. ISO uses the L_{nw} system [53], which is nearly identical save that the starting point is at 100 Hz and the endpoint at 3150 Hz.

3.4.7 Impact Sound Reduction

Everybody knows about the positive effect of a carpet on the impact sound reduction of a floor assembly. But is that really true? Let's create a simple experiment with a concrete floor on which a person is walking with standard shoes, and part of the floor is covered with a carpet. Should one go underneath and listen, it will be pretty obvious whether the walker is evolving on bare concrete or on carpet. Now let's have the walker remove his shoes and evolve again; this time it will be quite difficult for the listener underneath to decide whether the walker is located on the bare concrete or on the carpet.

The impact sound reduction coefficient of a floor covering is defined as the difference between the impact sounds measured in the receiving room of the laboratory with the impact machine located in the emission room on a reference floor (typically a 14 cm thick reinforced concrete slab) without and with, respectively, the floor covering under investigation. The measurement is typically carried out using standard ISO 140-8 [55] or ASTM E2179 [56].

3.4.8 Walking Noise of a Floor Covering

Nowadays one often is interested in the walking noise of the floor covering. This is assessed by performing inside the emission test room of the laboratory an impact sound measurement on a reference floor (typically a 14 cm thick reinforced concrete slab) with the floor covering under investigation. This can presently be carried out using EN standard 16205 [57]. The relevant quantity is written L_{new}.

3.4.9 Computation

A word of caution: Most of the time a crude model will not do, as the problem is more complicated than with sound reduction. For rough comparison purposes, one can use the rather crude expression

$$L_n = 133 - 30 \log(e) - 10 \log(V) - 3$$

where e is the thickness of the floor, expressed in cm, and V is the volume of the receiving room, expressed in m^3.

As of now, prediction models for the assessment of the impact noise of a floor have not yet been developed, especially when dealing with floors other than concrete. On the other hand, prediction models for the assessment of impact noise transmitted from one room to another are quite developed (they often are one option within a program that also deals with the sound insulation [15]).

3.4.10 Impact Noise on a Wall

So far in this book impacts have been considered as occurring on a floor. What about their occurring on a wall? This is not such a far-fetched problem, as one may face a ball being playfully bounced against a wall by a child, but also a noisy switch being fixed to a wall. More to the point, the acoustician usually looks at the impact sound performance of the floor prior to deciding on the noise reduction measures to be applied to the piece of equipment to be implemented on it, so a similar acoustic assessment of a wall on which a piece of equipment (e.g., a switch or a tap) will be installed is of interest. This is of particular importance when dealing with the noise generated by the operation of such a piece of equipment.

A small impact machine (known as pendulous hammer) has been developed for that matter and tested on various types of structures, including wooden structures [58].

3.4.11 Standards and Regulations

There are quite a number of standards and regulations pertaining to impact noise transmission (e.g., [21, 25–29]). A few of them are given in the various chapters of this book regarding the main types of construction. Typically, there often are maximum impact sound level values required for specific types of construction (e.g., dwellings, schools, etc.).

In addition, some sustainable development standards (e.g., HQE in France) require the floor covering to have been subjected to an impact sound measurement inside the test room.

The Swiss Engineers and Architects standard SIA 181 has taken provisions regarding the matter of noise induced by the manipulation of service equipment. A small impact machine (known as a pendulous hammer) has been standardized to assess the impact noise performance of the walls on which such equipment is mounted [59].

3.5 ACOUSTIC ABSORPTION AND REVERBERATION TIME

This chapter briefly outlines the basics of absorption and reverberation time. More explanations are given in Chapter 6.

3.5.1 Experiencing Reverberation

On entering a space, one has probably experienced a feeling like "it sounds dead" or "it sounds lively." This has often been a problem for performers when deciding on how much volume they should turn out, as different places of similar capacity may turn out to sound differently. More to the point, one has probably experienced trouble with the understanding of speech messages in a lively space. All these feelings are linked to the reverberation of the space (the longer the reverberation, the livelier the room will be perceived).

A professor, W. Sabine, found out that according to the number of cushions brought by his student inside the rather uncomfortable lecture theatre, the space sounded more or less lively. This eventually led to the notion of absorption quantity.

3.5.2 Acoustic Absorption

An absorptive material is a porous or fibrous material in which the vibrations of air are turned into heat when scrapping against the walls of the cavities. The efficiency of such a material is characterized by its absorption coefficient, theoretically ranging from 0 (reflective) to 1 (absorptive). In practice, the absorption coefficient value often appears as greater than 1 in commercial leaflets due to reverberant room measurement techniques (cf. Section 2.11.4.3). Measurements are carried out in a reverberant room (i.e., in a diffuse field where sound is supposed to be incident from all directions), typically on a 10 m² (according to standard ISO 354 [60]) or 6 m² (according to standard ASTM C423 [61]) room. It also is possible to assess the absorption coefficient under normal incidence using a wave tube according to ISO [62] or ASTM [63, 64] standards, but the relevant absorption coefficient value under normal incidence does differ from the one in the diffuse field, and picking up the pieces is not a task for beginners [65].

One may care to note that an absorptive material may be hidden by a cladding, but only as long as it is not airproofed. This is a significant difference compared to thermal insulation, where one can keep the thermal insulation performance using plasterboard in front of the insulation.

Often there is confusion caused by architects and contractors indifferently calling insulation a thermal insulation material or an absorptive material. While some materials actually manage to do both (e.g., a mineral wool), most of the time they are associated with a water barrier made of an airproof foil: This means they are no longer absorptive in the middle- and high-frequency range.

3.5.3 Reverberation Time

Reverberation time (RT) is a bit similar to the time constant of a physical system: The longer the time value, the greater the stability, but on the other hand, the slower the reaction to a change of excitation. One can experience it when practicing music: In a large church one can hear the organ sounding for several seconds after the key has been turned off (but one will also experience trouble understanding the articulation of consonants in such a space); in a small upholstered meeting room there will not be any hope of sustaining sound (but one will also easily understand the articulation of consonants in such a space). This hints at two important factors: volume (the greater the volume, the longer the RT) and acoustic absorption (the smaller the absorption, the longer the RT).

The reverberation time is defined as the time span needed for the sound level to decrease by 60 dB after the sound source has been cut off. Another quantity, named early decay time (EDT), deals with the first 10 dB of decrease.

More RT-based indicators are discussed in Chapter 6, which is devoted to room acoustics.

3.5.4 Spatial Sound Level Decay

Reverberation time is not always suited as a descriptor of the internal acoustics of an enclosed space; for example, an encumbered flat space with a reflective low ceiling may feature a reverberation time well under the 1 s mark (due to diffusion effects on the fittings), while its users will find it uncomfortable due to sound propagation by means of reflections on the ceiling. This is a kind of situation encountered in open-space offices or industrial spaces. To cope with the task of describing the acoustics of such spaces, one uses the notion of spatial sound level decay. When measuring away from an omnidirectional sound source the sound level value first decreases rather sharply with the distance, as the direct sound field of the sound source is predominant. Next, the reverberant sound field is an important contributor (intermediate region), and then later (far field), it is much greater than the direct field, as illustrated in Figure 3.3.

The spatial sound level decay (DL_2) is defined as the rate of decay per doubling of distance. It is measured in the intermediate region according to standard ISO 14257 [66] using an omnidirectional sound source. Prior to the introduction of this quantity, a French law text had used it for industrial spaces, either fitted or not [67]. Lately, ISO 3382-3 has proposed a measurement procedure for open-space offices [68]. The result may be expressed as a single number quantity for either a pink noise or a speech noise.

Note: For quicker assessment (or when the dimensions of the room are not compatible with a spatial sound level decay measurement!), one sometimes uses the so-called amplification, which is the difference between the sound pressure level measured at 10 m from an omnidirectional sound source in the space under study and the sound pressure level measured at 10 m from the same sound source under free field conditions.

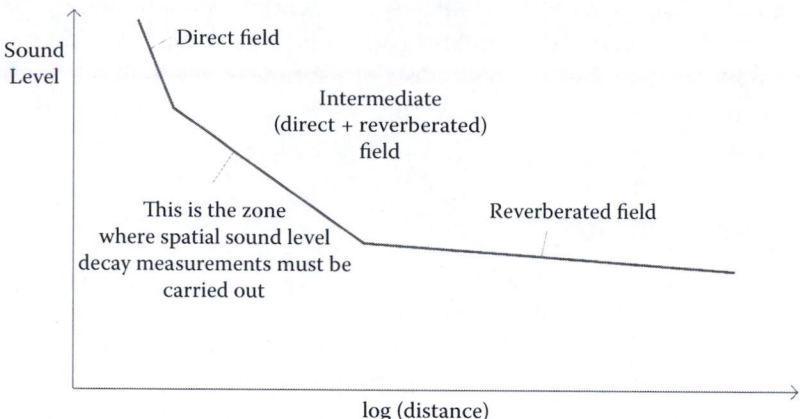

Figure 3.3 Spatial sound level decay.

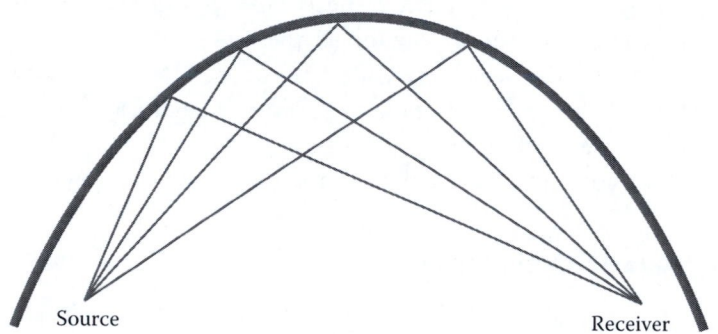

Figure 3.4 Example of focusing due to a concave surface.

3.5.5 Reflecting, Focusing, and Scattering

So far we have not bothered with the shape of the room. Unfortunately, it will play a role. To start with, a concave surface will usually tend to focus sound energy as illustrated in Figure 3.4. While this can be exploited for some effects, it often proves to be a hindrance. That can be prevented using either an absorptive material (that will absorb sound energy) or a scattering material (that will redistribute the sound energy over a large area instead of a specific direction) (cf. Section 6.6).

Note: Should the relevant acoustic treatment be mounted on the wall as a separate fitting, please do keep in mind that its mounting may affect the overall sound reduction index (e.g., due to the small air gap between fitting and structural wall, or simply due to the fixation holes!).

3.5.6 Standards and Regulations

There are a few standards and regulations pertaining to reverberation time and equivalent absorptive area [69]. A few of them are given in the various chapters of this book regarding the main types of constructions. Typically, one may care to note the following constraints:

- There are limits, expressed as noise levels, regarding the noise exposure of workers, or the danger limit for people exposed to high-level music. Introducing absorption in the room may help reduce noise propagation and sound level value, and it is often recommended in the relevant guidelines.

- There often are maximum reverberation time values required for specific types of construction (e.g., dwellings, schools, etc.), as well as minimum equivalent absorptive areas for such types.

3.6 VIBRATION CONTROL

3.6.1 Foreword

Building vibrations is a domain of its own. In this chapter we will limit our interest in such vibrations that can actually radiate audible or perceptible noise.

3.6.2 Sources of Vibrations

Vibrations in a building can be generated by various vibratory sources. Here are a few of them:

- Mechanical equipment
- Users' equipment
- Rail transportation corridor nearby
- People walking or dancing
- Wind on the façade or roof

3.6.3 A Few Standards and Regulations

While standards and regulations pertaining to noise exposure have steadily been developed over the years, texts pertaining to vibration exposure are not as obvious. This is in part due to the lack of data on the subject. While aspects covering excessive vibration levels are usually covered due to the risks to the structure of the building, aspects covering annoyance are not as completely covered. The international standard ISO 2631 [70] considers three vibration spectral limits, for comfort, work efficiency, and danger, respectively, with the highest sensitivity in the 4 to 8 Hz interval.

When it comes to noise generated by vibrations, there are regulations in a few countries [71]. Those typically set a maximum L_{Aeq} value over a given time span. For example, a Swiss federal regulation [72] requires the $L_{Aeq,1h}$ generated by rail transport not to exceed 30 dB(A) in nighttime.

3.6.4 Vibration Control

Vibration control philosophy is a bit similar to noise control, as one can act at three different stages:

- Reduction at the source (i.e., minimizing the amount of vibratory energy generated by a vibratory source through either better design or resilient decoupling from the structure)
- Reduction along the path (i.e., implementing expansion joints along the propagation path)
- Reduction at the receiver end (i.e., box-in-box construction)

3.6.4.1 Reduction at the Source

Reduction at the source of course happens to be the most efficient way to deal with a problem of vibration-induced noise, as it will prevent spreading vibratory energy throughout the building. Now for the bad news: it will also not be possible under any circumstances, as it will certainly entail an interruption of the operation of the equipment involved in order to implement the required measures. This is especially true of rehabilitation projects. For example, when dealing with a freezer serving a whole building, the users may not be happy at the length of time required to disconnect the equipment, build the resilient supporting elements, reconnect everything, and test it prior to recommissioning. Even worse, when dealing with a railway line, there usually is no hope of implementing vibration control measures without a significant interruption. When small gains (up to 5 dB at 63 Hz) are looked for, there is a technique for light rail systems using a resilient fastening of the track on the sleepers. As the sleepers do not need to be removed from the track bed, this does not entail speed reductions, and it can be implemented at a suitable time, for example, during the night break of operations [73].

3.6.4.2 Reduction along the Path

Reduction along the path is tricky, to say the least. In aerial acoustics over distances under the 100 m mark one faces the same propagation medium. When dealing with vibrations in the ground, the propagation medium can be rock somewhere and sand somewhere else, not to mention all those forgotten structures left interred. This may sometimes give surprising results when it comes to the propagation in the ground. One cannot make the economy of a diagnosis. This is made through vibration measurements at given distances from the source, and whenever possible, the soil test excavations are used too.

If the vibrations are mainly propagated on the surface, it may be expedient to try to use a vibration barrier [74]. This is efficient as long as no objects or harder ground are located nearby and likely to reflect energy short-circuiting the barrier.

It is also often possible to implement vibration control measures between the foundations and the superstructure of the building. This is performed using springs or resilient elements that are typically located on the top of the foundation pillars or on top of the basement structure. Close cooperation is needed between the structural engineer (as the performance of the springs or resilient elements will depend on the accuracy of the load applied on those elements), the acoustic engineer, and the safety engineer (as consequences of a fire or an earthquake must be investigated).

3.6.4.3 Reduction at the Receiver End

As a last-ditch effort, vibration reduction measures can be applied at the receiver end.

Those will typically entail a box-in-box construction, with the rooms to be protected built using a concrete slab supported by resilient pads or springs over a stiff concrete structure. Of course, such a scheme does complicate the actual construction work.

3.6.5 Noise Generated by Vibrations

Vibrating surfaces will radiate a sound power level. Usually this is in the low-frequency range. For example, inhabitants of downtown Parisian dwellings are quite accustomed to the deep grumble (63 Hz) from the underground rail lines. Noise radiation will actually appear well before any vibratory sensation.

The sound power level L_w radiated by a wall of surface S excited by vibrations can be expressed as

$$L_w = 10 \log(S) + L_v + 10 \log(\sigma) + K$$

where σ is the radiation factor of the wall, L_v is the velocity level reference 10^{-6} m/s, and K is a constant

The radiation factor s is equal to:

1 if $f > f_g$

$0.45 \, \mathrm{sqr}(P/\lambda_g)$ if $f = f_g$

$1/\pi^2 \, (P \, \lambda_g/S) \, \mathrm{sqr}(f/f_g)$ if $f < f_g$

where P is the perimeter of the surface.

The coincidence frequency f_g is given by

$$f_g = c^2/(2\pi d) \, \mathrm{sqr}[12 \, \rho \, (1 - \mu^2)/E]$$

where E is the Young's modulus (in N/m²), d is the thickness of the plate (in m), ρ is the volumetric mass of the material, and μ is the transversal compressibility coefficient.

Here are a few examples of $10 \log(\sigma)$ values [75]:

For a 24 cm brick wall: 0 dB.
For a 7 cm concrete: Range −15 to −5 dB until 500 Hz, then 0 dB above.
For a 13 mm plasterboard: Range −15 to −5 dB until 2000 Hz, then 0 to 5 dB above.

3.7 CONSTRUCTION NOISE

There usually are a few rules pertaining to the construction. In order to reduce annoyance to the neighborhood, there will typically be a specific time span to be complied with (e.g., 7:00 a.m. to 7:00 p.m. at most), with night work and weekend work allowed on a case-by-case basis.

Most of the time, the core of the requirement is contained in the building permit, that states the allowable hours of operation and the eventual restrictions (e.g., more stringent hours and sometimes the allowed methodology of demolition and construction) [77–79].

When the building site is located in a sensitive environment, it is not uncommon to have a noise monitoring system installed and regular measurements performed.

In some situations the building permit can be a real killer. As an example, in Paris on the famed Champs Elysées, the regulations in force require the waste bin to be brought on the curb at 6:00 a.m., and it must be removed no later than 7:00 a.m.! Needless to say, the squeaking and banging noises from these operations are not pleasant to the neighborhood.

Word of Advice: While there are a few legal requirements in force, it is not a bad idea to try to contact the neighborhood to explain what the building operation will be and when a few noisy phases will occur [77, 80].

3.8 A FEW STAGES OF BUILDING CONSTRUCTION

A building project is typically carried out over several phases (please note that in smaller projects some of those phases may be grouped together!). Here is a brief description of those phases.

Sketch, feasibility sketch: Analysis of the program drafted by the end user, draft sketches, technical and architectural notice, preliminary economic estimate. Note: This is the right moment to point out the potential requirements and problems of a project, for example, try to get a sensible layout of the rooms inside the building with regards to one another and with regards to the potential exterior environment.

Preliminary design: Analysis of the program (complements if need be), checking that the sketches are compatible with the technical and architectural requirements, technical notices, architectural notice, cost estimation, general drawings. Note: This is the right moment to ask for specific constructive elements (e.g., a plain concrete slab, an acoustic door, etc.).

Design development: Verification of compliance with the regulations in force, adaptations and final complements of the program, finishing touches to the administrative and technical documents, drawings of floors and cross sections, architectural details, prescriptions regarding architectural and technical items, final cost estimate. Note: This is the time to be committed to the budget (do make sure nothing was forgotten!).

Final development: Analysis of the choices made during design development, detailed notices and drawings. Note: This is the right time to call for suitable sealing around the various networks or constructive element; this is also a suitable time to require acoustic testing of some assemblies.

Tender document: Elements to be addressed by the contractor in the submission, using the final development document set. Note: This is the right time to require from the contractor a specific methodology or a list of conditions to be able to properly check the work.

Assistance to contractor selection. Note: This is the right time to make sure one will not be plagued with a contractor who was the lowest bidder but who is unable to properly perform the required work!

Visa: Visa of execution drawings and equipment selection, choosing architectural and technical solutions on the basis of the synthesis of the project. Note: This is the right time to prevent a problem through the erroneous selection of unsuitable material or equipment by the contractor.

Site supervision: Participation in the organization of the work, checking that work is proceeding according to the requirements. Note: In France one says *"pas vu pas pris"* (not seen not catched). One definitely must be able to check such important details as, for example, the absence of elements likely to short-circuit resilient fasteners or pads, the use of a concrete or plaster finish instead of a glued plasterboard, etc.

Commissioning: Assistance to the complete commissioning through measurements. Note: This is the right time to emit any statement regarding the noncompliance of the contractor's work with the acoustician's specifications. This is also the right time to explain to the end user or his representative how to use the facility that is being delivered.

By the way, a project usually requires some demolition work at the beginning of site work. Whenever this is carried out with some thoughts for sustainability, the politically correct designation of this phase is deconstruction.

3.9 REFURBISHMENT

Rehabilitation is a quite usual procedure in Europe [81]. There may be quite a few reasons to go for it. To start with, older buildings usually feature a higher legal potential occupation ratio per ground square meter than the ones allowed by normal recent constructions. More to the point, it helps the urban planners to keep a homogeneous urban appearance. Last but not least, it helps save time if the walls and floors (and sometimes even the roof) are kept; administratively speaking, it may also speed up things, for when the envelope of the building is kept, one often will solely require a fitting-out permit instead of a full building permit. On the negative side, it is not uncommon to lack some suitable space to route the ducts and pipes through. Quite often, the walls and floors do not feature high enough an acoustic performance and must be completed using plasterboard and mineral wool stud-mounted elements that will reduce the available floor dimensions and height.

In such a kind of project it is necessary first to perform a diagnosis of the existing building in order to find out how it is built and where the sensible points are. Each specialty will have to perform its own diagnosis. One should remember that a structural diagnosis usually can be a bit destructive, as the structural engineer will typically cut through floors and walls in order to assess their composition, so the acoustician must make his measurements before.

Note: There is a strong need for coordination between the interested parties, as depending on the planned sketch of the future rooms inside the building, some zones will be more sensible than others. More to the point, everybody must be aware of each other's needs.

The diagnosis through measurements will feature:

- Sound insulation measurements between rooms when the walls are kept. Note: This measurement will help assess the potential flanking transmission by those walls, as well as the potential sound insulation of those rooms.
- Sound insulation measurements and impact sound measurements between rooms at different floors. Note: This measurement will help assess the potential sound insulation between floors.
- If applicable, vibration measurements on the floors. Note: This will help assess the eventual noise generated by vibrations (e.g., from rail lines nearby) and vibration levels inside the building.
- An acoustic diagnosis of the site will be performed as per a regular new construction project (i.e., assessing the sound level values on the site and finding out what the potential noise sources around are, as well as the potentially sensible zones around). Note: Do keep in mind that usually a rehabilitation project will entail some unbuilding prior to the actual construction work. Under the nice words one can already hear the concrete breakers hammering away, so one had better have a good look at the location of the nearest neighbors, especially those who are structurally linked to the building. It will probably be necessary to explain to the neighbors the basics of the project and point out that while some phases of the work will be noisy, they will be kept to a minimum of duration and their time schedule will be adapted, while appropriate noise reduction measures will be implemented.

It must be stressed that the diagnosis will constitute the testimony to the acoustic performance of the building prior to any work. It is not only a basis for the acoustic studies (from which predictive computations will be elaborated), but it is also often a compulsory step to be able to prove ultimately that the initial acoustic performances of the building have not been deteriorated [81].

In the particular case of historical buildings things can get quite complicated, as usually the façades and even the roofs must be preserved. In some cases, it is even necessary to preserve some interior spaces (e.g., because of paintings on the walls or ceilings). Under such circumstances the acoustic objectives must be adjusted on a case-by-case basis, and specific solutions must be elaborated (e.g., introducing intermediate spaces around in order to prevent direct transmission to other spaces of interest, or working on the other side of the partition or floor using such doublings as a floating floor, a plasterboard ceiling, or half wall, with mineral wool in the void).

A special mention must be made regarding performance halls: Those usually are considered historical landmarks, and the end user may wish to preserve (or sometimes improve) much more than the sound insulation characteristics (cf. Chapter 6).

3.10 NOTIONS OF SOUND MASKING

3.10.1 Foreword

Sound masking has enjoyed a few developments due to the increased use of electronics, but it is by no means a recent method. In antiquated times, the Romans would often use the sound generated by a small fountain in the atrium to try to mask noise from one room to another.

Nowadays, everybody has had a few occasions to wonder about the sound insulation of one's room. Actually, the perceived intrusion of noise is only one of a few contributors to the global noise level in the room (cf. Figure 3.5). When those contributions are well balanced, usually one does not especially overreact to the various sound stimuli, but if one of those contributions is much greater than the others, then you can bet that some noise annoyance will soon be felt by the inhabitants. A very classic example of sound masking is that of the rehabilitation of dwellings: In a first phase, an acoustician will be required to devise a significant improvement of the façade insulation; within a year, he or she will come back and ask for an improvement of the sound insulation between flats. What is happening? Prior to the rehabilitation the noise transmitted from outdoors through the façade and the noise from other flats were more or less balanced; when the façade sound insulation was improved, then

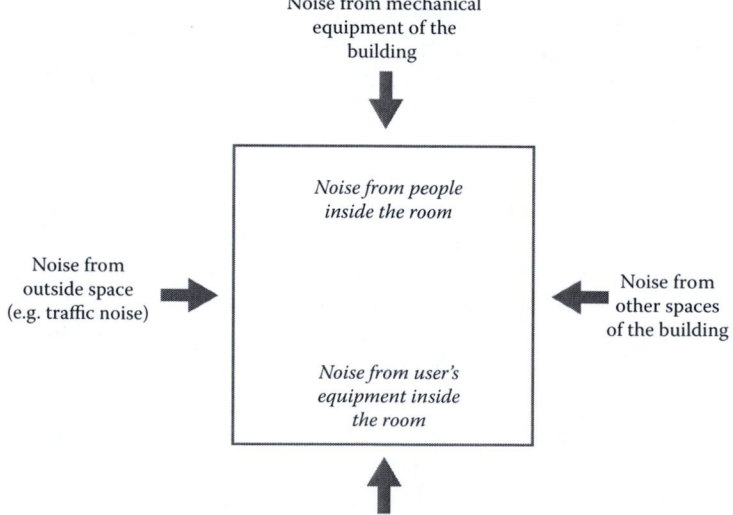

Figure 3.5 Noise contributions in a room.

the background noise level in the flats was significantly reduced and any noise from other flats was plain to hear and identify.

3.10.2 Sound Masking

As implied by the name, sound masking does not eradicate a particular noise signal (known hereafter as a distracting signal); it merely tends to hide it from the potential listener. So, one has to use a sound system that will raise the background noise level value above the sound level value of the distracting signal (e.g., a voice). Of course, such an increase cannot be performed with too high a sound level (e.g., over the 45 dB(A) mark the sound masking signal will start being perceived as another distracting signal too). More to the point, the spectral contents must cover those of the distracting signal. Last, in order to avoid being distracting, the masking sound usually is steady.

Adjustments to both noise level and spectral content should be left to the specialist (actually, one of the first technical ventures in this field featured loudspeakers on top of open-plan office furniture panels that unfortunately had controls readily accessible to the user, and that spelt its demise). Nowadays, systems are implemented using a computer simulation, and their loudspeakers are often also used for such purposes as background music or PA messages.

Please do remember that the masking sound must not be recognizable by the user, as it would then become a further distraction in the soundscape, and it could even become one more nuisance. For this reason one must refrain from using music as a masking sound (while it might be enjoyed by some people, it will annoy others).

Examples of use are open-plan offices, waiting rooms, security-minded offices, etc.

3.11 A WORD ABOUT PRIVACY AND SECURITY

3.11.1 Foreword

Everyone has been subjected to the acoustics of offices or dwellings with varying degrees of sound insulation performance. One has probably noticed that when the background noise level value is especially low, it becomes easier to hear what is going on next door. One may also have found that in flats during the daytime, one merely feels a presence in the next flat, while at nighttime it may be possible to recognize voices.

Actually, the ability to hear and understand what is being said in a room depends on the intelligibility of the message (more about that will be covered in Section 6.5). This means that on the receiving side it will mainly depend on the sound insulation between the rooms of interest and on the background noise level at the receiving end.

As a rule of thumb, the minimal value of the sound insulation D_{nTw} needed to ensure privacy is 45 dB.

3.11.2 Privacy and Security

Privacy means that under normal circumstances (i.e., a normal spoken voice on the emitting side, and a background noise on the receiving side featuring the usual mechanical services and small chatter), one will not understand what is being said at a normal voice level on the emitting side. This will entail a minimum 45 dB sound insulation. But should the voice level be raised or the background noise level drop, this will be an entirely different matter.

If one wants to avoid such a situation, such contingencies have to be taken into account, and a sound insulation of 55 dB is a minimum.

But there is worse to come: Should somebody be intent on eavesdropping, there are quite a few tricks much more sophisticated than the old ear on the partition, that is, a vibration captor on the wall or floor (cf. Section 3.12.17), or even measuring the vibration of such a surface using laser equipment.

Water-filled ducts usually are poor acoustic conductive paths. But empty or air-filed ducts are another matter, and it may be necessary for security purposes to fit them with a vibrator too.

Masking sound comes handy in such circumstances. For simple cases, one will merely have a sound system adjusted so as to ensure that the background noise levels around the protected room are more than 10 dB above the transmitted speech sound. For more complicated cases, one will also introduce vibration generators on the walls and floors in order to disrupt any attempt at measuring the vibrations of the envelope of the room.

3.12 EXAMPLES

3.12.1 Is Such a Heavy Concrete Wall Insulating?

When they had to prescribe the means of sound insulation between two rooms, both the small contractor and the end user of a small extension project decided a heavy-cast concrete wall would do fine. However, when the time for trials came, both were startled to discover that some rather high-pitched transmitted noise could easily be heard.

Looking closely at the work, an acoustician invited to have a look quietly pointed out that the cause was to be found in the improperly filled holes in the wall that previously connected both sides of the casting frame.

Lesson Learned: The overall acoustical performance is dependent not only on the chosen material, but also on the modus operandi.

3.12.2 A Small Dance Floor

A small dance room (capacity: 10 persons maximum) had been fitted in a public building in what formerly used to be a small storage space. The floor covering was of a PVC type.

When people were dancing on that floor, the lighting fixture in the service dwelling underneath made strong oscillations. More to the point, the dancers were complaining of the uncomfortable conditions on such a floor (Let's face it: The floor was too stiff for the dancers but too weak for acoustic purposes).

The whole floor assembly was eventually replaced by a thick concrete slab and a wooden floating floor on resilient joists, which solved the problem.

3.12.3 Impact Noise Reduction

The new owner of an old flat decided that it would be nice to get rid of the tired old-looking carpet in the living room. He decided to have it replaced by marble tiles. In order to reduce impact sound transmission to other parts of the building, it was decided to lay them on a resilient screed. As soon as the work was finished, the neighbor underneath complained of too high impact sounds. It turned out from the measurements that while the impact noise levels did comply with the legal objectives applicable for new construction, they were slightly greater than those measured with the old carpet. The owner eventually managed to get an agreement with the condominium to keep his new floor covering, as internal rules had provisions for such a case.

Lesson Learned: It may not be sufficient to look at the legal requirements; degrading the acoustic quality may often be considered a problem.

3.12.4 A Fan

The acoustician had been called in a flat where the tenant was experiencing a strange sensation. On arriving there he found that there were strong stationary waves in the 16 Hz third octave band in the bedroom: The noise levels measured close to the walls and in the middle of the room were higher by 10 dB than the other levels. It eventually simply turned out that the tenant at the previous floor had mounted an unbalanced ceiling fan.

3.12.5 Misusing a Computer Model

One day the supervisor of a public facility building site phoned the acoustician and expressed his dissatisfaction with the prescriptions: All required wall thicknesses were wrong, e.g., a 12 cm concrete wall was sufficient where the acoustician had required a full 20 cm wall.

It turned out that the supervisor had used a commercial computer model that was suitable for dwellings. When applied to large volumes such as a conference room, the correction for the depth of the receiving room was such that it would theoretically enable the use of much smaller separating walls.

In another example, there was a project of offices being built close to a very busy motorway. While the acoustician prescribed the façade characteristics bearing in mind that there would be partitioned offices along the façade, the contractor considered the whole open-space floor. This resulted in a 10 dB difference on the sound reduction index of the façade.

3.12.6 A Window

A manufacturer writes in its technical data that the sound reduction index of its window is 33(1; −4). What does that mean?

Well, this is a single number according to ISO 717-1, stating the R_w value and the corrective terms to access the pink noise and the road traffic noise sound reduction index values.

This gives

$$R_w = 33 \text{ dB}, R_A = R_w + C = 34 \text{ dB, and } R_{Atr} = R_w + C_{tr} = 29 \text{ dB}$$

Lesson Learned: Be careful should one be dealing with specifications involving a road traffic noise reduction index, as most distributors and contractors are not always aware of such subtleties!

3.12.7 A Glued Plasterboard

When using concrete blocks, the architect often wants a proper finish applied to the wall. In order to improve its sound reduction performance, the acoustician usually wants a 2 cm minimum plaster or concrete finish to be applied. However, due to the task being quite time-consuming, it is quite tempting for the contractor to use glued plasterboard.

The problem is one will end up with a 1 cm gap between the structural wall and the plaster-board, and more to the point, the concrete blocks will remain as porous as ever. In the end, one will discover that while the general visual aspect is satisfying, the acoustic target is not met due to a strong resonance.

3.12.8 A Glued Calibel (Plaster on Mineral Wool)

Calibel is a complex made of a 50 mm glass wool factory assembled onto a 10 mm plaster-board. This is a kind of product often used by patrons to try to improve the sound insulation of their premises. However, due to its dimensions, while it can be effective for speech transmission prevention, it can actually degrade the sound reduction performance in the low-frequency range (e.g., when attempting to reduce the noise from the television set next door).

In a particular case when the acoustician had been attempting to reduce the noise of a washing machine transmitted through a light wall from the kitchen next door, the contractor decided to use Calibel instead of plasterboard on metal studs. The result was clear enough, as the measurements actually showed that the final situation was worse than before by 3 dB due to the low-frequency range transmissions.

In another example the contractor found it more expedient to use a Calibel instead of a stud-mounted plasterboard doubling that had been required to improve privacy between two rooms separated by a light partition. The result was startling: With the glued complex added on the original partition, the sound insulation had been reduced by 3 dB at 500 Hz.

Lesson Learned: Do not trade a stud-mounted doubling for a glued doubling.

3.12.9 Long Reverberation Time in an Anechoic Room?

When commissioning its new anechoic room a university laboratory decided to have students perform reverberation measurements inside. To the consternation of the professor, several students reported a quite long reverberation time in the 4000 Hz third octave band. After giving them a proper dressing down (how could one expect a long reverberation time with so much acoustic absorption around?), he ordered them to do a proper measurement. Back they came with similar results. The baffled professor then personally performed the measurement and got the same results too. It eventually turned out that rodents had nested in the mineral wool and screamed at 4000 Hz whenever they heard such a sound.

Lesson Learned: You may believe your measurements, but only if you can explain them!

3.12.10 Ruined Absorption

A project had required some absorption in the room, and the required amount had been provided using a perforated plasterboard ceiling. Unfortunately, the reduced time frame drove the architect to order the painters to proceed hastily. When this was finished, all the holes and porosities had been clogged, turning the previously absorptive ceiling into a reflective one.

In another case, a laboratory had enjoyed a poor man's semianechoic facility made of a soundproofed cabin featuring mineral wool protected by perforated metal plates inside. One day an official came visiting the lab and decided a repaint was in order. By the time the lab technician realized what was going on, his semianechoic test room had been turned into a fairly reverberant room!

Lesson Learned: Always make sure the reasons for such materials have been explained.

3.12.11 An Improvement Turned Sour

An acoustician got a phone call that left him with a flea in his ear: An architect told him that his absorptive treatment prescription for a small performance hall was an absolute disgrace. Puzzled about this project that did not ring any bell, the acoustician eventually found out that this particular architect had simply copied the text from the prescription book of another of the acoustician's projects; in doing so, he thought he had made a little improvement by replacing this itchy yellow or pink material by friendlier polystyrene!

Lesson Learned (at great cost!): Do not copy! More to the point, an absorptive material features an airflow resistivity, if you know that you won't use polystyrene as an absorptive!

3.12.12 A Bridge

Here are a couple of examples of bridge vibration events (and kindly remember, some constructions are built as a bridge!).

In the mid-19th century a troop marched on a river bridge in France. According to some stories, the rhythmic pace of the soldiers was close to the resonance frequency and started to excite the bridge, and this was considered much fun—until the bridge collapsed because of it. A similar story was reported in Great Britain for the Broughton Bridge in 1833.

Another famous story concerns the Tacoma Bridge in the United States that was excited by the wind in 1940. The excitation eventually reached such an intensity that the bridge broke. There was a film capturing the whole sequence [76].

3.12.13 Structure-Borne Rail Noise

Here are a couple of examples of structure-borne rail noise control attempts.

Foreign investors bought a Parisian office building also housing cinemas and shops with the idea of turning it into a luxury hotel. When they realized that noise was generated in the cinema facility due to an express rail line and a subway line (which also had the bad grace to be located at the ideal spot for the parking ramp), they simply went to the transportation authorities to ask for those lines to be deviated (quite simple, isn't it?). Of course, the bewildered authorities turned the request down. That project eventually featured projection theatres built as a box in box.

Lesson Learned: Do not even dream of deviating a rail line!

In another project, in order to achieve the required structure-borne noise reduction in the office building above a railway tunnel, the tracks from the suburban rail line were subjected to vibration control measures using a resilient mat under the ballast. Such a move was only made possible by the closure of the line during 1 month due to track and infrastructure maintenance.

Lesson Learned: At the source vibration control may be possible with proper planning.

In another rehabilitation project it was wondered whether it would be possible to reduce the structure-borne noise from the subway inside the basement of a historical building. Investigations quickly showed that there was a maze of sewers, ducts, and technical passages between the walls of the subway tunnel and the foundations of the building. More to the point, part of the tracks had already been treated to a resilient mat under ballast. With propagation paths coming from practically everywhere, it was found totally uneconomical to try to create a clear resilient separation between the foundations of the building and the tracks, and the idea of structure-borne reduction was dropped from this project.

3.12.14 Shop Offices and Dwellings

An art deco building located on the famed Champs Elysées in Paris housed a luxury shop, as well as offices and dwellings. The owner decided a major rehabilitation would eventually have benefits through a higher rental value, and a project was decided upon.

The diagnosis was performed by the whole design team. It showed that while the sound insulation was quite good at the lower floors with a 53 dB value between floors (that usually were the master's apartments in the olden days), it steadily deteriorated, with only 38 dB between the last floors (that were the servant's quarters). It also showed that the structural capabilities of the building were quite limited.

The project called for plasterboard walls and ceilings in order to save weight, while increasing the sound insulation between spaces of the building. Resilient floor coverings were systematically used. One of the difficulties was to find the necessary space to route the new ducting system and install the air handling units, with some of them ending up in a former servant's room and the others in the basement.

The deconstruction planning and the building planning had to be carefully studied, as the shop at ground level was operating in the end of the afternoon, and there was a hotel next door too. Discussions with their respective operators enabled to them reach an agreement through which the building contractor could perform normal work on weekdays between 10:00 a.m. and 3:00 p.m. The methodology of work was adapted to the circumstances; e.g., sawing under an absorptive enclosure was preferred to the use of a concrete breaker. The only significantly annoying point was the waste removal, as local regulations prescribe the bin to be brought no sooner than 6:00 a.m. and to be removed no later than 7:00 a.m.!

Lesson Learned: Rehabilitation work is possible in a dense urban environment as long as appropriate measures are taken and discussion is kept opened with the neighbors.

3.12.15 Where Is One Supposed to Measure?

An expert to the court was required to perform some noise level measurements in a bedroom of a multidwelling building where the tenant complained about various noises coming from other parts of the building. When he came to perform the measurements he was presented with an empty room where he performed the usual sound level measurements at a height of 1.5 m from the floor. No significant noise event could be identified.

Then it suddenly hit him: If this was supposed to be a bedroom, where was the bed? The tenant looked blankly at him and opened a small cupboard from where he picked up his braid (well yes, he was of Asiatic descent)! When measuring close to the ground, a number of events were at once identified.

Lesson Learned: When somebody complains, it is interesting to perform a measurement at their usual location and not limit oneself to the typical locations indicated in the standard.

3.12.16 Is There an Insulation Problem?

On performing the sound insulation measurements between two cinema projection theatres, the acoustician felt quite confident as he had already designed and built similar facilities along the same design lines. In this particular case, the separating wall between those halls was made of a double concrete wall with an expansion joint in between that should ensure the required performance. So sure was he that he sent a young engineer to perform the measurement while he talked with the end user. To complicate matters, the brand new amplifiers were smelling a bit too hot for comfort, and it was decided that one should not push the volume too far for the sake of safety of the sound equipment (well, it wouldn't have done to burn the amplifiers before the official inauguration by the authorities, would it?). While this was supposed to be a bit close to the limit in the high-frequency range, it was supposed to be usable. The young engineer came back and stated that he felt the measurements were OK.

To everybody's surprise the sound insulation result turned out to be rather bad, so bad that spectators complained soon after the opening of the facility. On coming back, the acoustician had the sound system turned on in earnest. While a low-frequency rumble could be heard on the upper part of the wall, a rather high-pitched noise could be heard close to the screen.

When the acoustician pointed out that such a noise could only come from a hole in the walls, the architect was adamant that it was not possible, as nobody in his right mind would have drilled a hole through two concrete walls, and it would be luck that they would match each other. But the acoustician held his ground: This did sound like a hole. Well, exploring the wall with his hand, the end user suddenly found a very soft spot and the noise was suddenly considerably reduced. Such a double hole had really been drilled (for which purpose nobody knew!). While he was wondering why his young colleague had not identified the problem, the end user came up with an explanation that was duly tested and proved conclusive: With a lower sound level on the emission side, it was really much harder to distinguish the noise radiated by the hole.

That did solve the high-frequency part of the problem (and a good thing that was, as the speech signal in a projection theatre could then be heard in the other theatre because of it). Regarding the low-frequency rumble, it eventually turned out that the contractor had left in place the material used to cast the double concrete wall, and this created a surface coupling that significantly decreased the low-frequency sound insulation performance.

Lessons Learned: Use a high sound level pink noise on the emission side to be able to spot weak spots. Do not rely on a young unsupervised collaborator to make the assessment, and do not take for granted that something will not be done because it is not logical (it might not be at the time of analysis, but it briefly made sense to the contractor for reasons of his own).

3.12.17 Listening to Conversations through a Floor

A large rehabilitation project involved the demolition of a few concrete structural elements. In order to ascertain whether some parts of the building could still be kept in operation during the deconstruction phase, it was decided to perform a diagnosis on one unoccupied floor. It involved banging on the floor and walls of a room at one end of the building and checking how much sound energy was lost at each expansion joint using vibration measurements.

When the acoustician analyzed the tapes, he found that after the second expansion joint, the banging was no longer much discernible. On the other hand, the conversations held in the room underneath were totally understandable!

Lesson Learned: A properly executed expansion joint is an efficient attenuator; also, eavesdropping using accelerometer measurements is possible.

3.13 HAVING A GO AT DIMENSIONING THE PROJECT

3.13.1 General

The first step will be an inventory of the applicable regulations and standards: They probably feature noise limits that have to be complied with, as well as some other acoustic targets (e.g., reverberation time and sound insulation requirements). In addition, there often is a program that has been drafted for this specific project, and its targets must be checked against legal requirements as well as against technically and economically realistic customs.

The next step will be an inventory of all potential noise sources (i.e., determining what the noise levels are likely to be in the various spaces of the project, including its environment). Such data may be found in the program established by the end user's representative; should that be not the case, a hypothesis will have to be stated and identified as such.

There will also be an inventory of all sensible spaces (e.g., bedrooms, etc.) to be found in the project. The relevant noise limit will have to be determined. Such data may be found in the program established by the end user's representative; should it be not the case, a hypothesis will have to be stated and identified as such.

Prior to actually dimensioning the various constructive elements, please kindly remember that:

- The sound reduction index of a wall or floor, and the impact noise level of a floor, measured under laboratory conditions, usually are officially given with ±2 dB accuracy; in practice, it is closer to ±5 dB, and one must take such a value into account to be on the safe side when performing the computations.
- Absorption coefficients are often given with 20% accuracy.
- Sound power levels of a piece of equipment usually are officially given with ±2 dB accuracy; unfortunately, they are often given under operating conditions that may seriously differ from the actual conditions to be encountered in the project.

3.13.2 Sound Insulation

Next, one will have to state the sound insulation objectives required for the compliance with the noise limits stated for each space. One may care to remember that the global sound level inside a space is the sum of several contributions (cf. Figure 3.2).

Last, one will have to check that the above-computed sound insulation values are not smaller than the eventual values that are required by the regulations in force. A similar check will have to be performed with regards to the values required in the program.

Those requirements will result in minimal specifications for the envelope of the spaces (floors, walls, etc.).

3.13.3 Impact Sound Insulation

One will have to state the impact sound insulation objectives required for each space.

Next, one will have to check that the above-defined specifications in Section 3.13.2 still enable the project to reach the impact sound insulation targets. In the negative, extra prescriptions will then apply (e.g., definition of a resilient floor covering or even a floating slab).

3.13.4 Reverberation Control

One will have to state the reverberation time for each space.

The reverberation time may be estimated using Sabine's formula. One will have to check that the computed values are not greater that the eventual values that are required by the regulations in force. A similar check will have to be performed with regards to the values required in the program.

3.13.5 Mechanical Noise Control

The initial sound level objectives were stated in Section 3.13.2. Now one will have to state the sound level contribution of the mechanical equipment required for the compliance with the noise limits stated for each space. A preliminary computation will have to be carried out, from the circulator (e.g., an air handling unit (AHU)) to the delivery point of interest (e.g., a louver in a room).

In addition, one will have to make sure that the structure is capable of accepting the weight of the equipment under scrutiny, plus its attending inertia mass if required. More to the point, one will have to make sure that there still is enough headroom left according to both work condition regulations and manufacturer's specifications.

3.13.6 Fire and Safety

One will have to check (or have a specialist check) that the various assemblies of constructive elements comply with the fire and safety regulations in force. Should that be not the case, it is mandatory that the prescriptions be revised accordingly.

REFERENCES

1. ISO 140-3: *Acoustics—Measurement of sound insulation in buildings and of building elements—Part 3: Laboratory measurements of airborne sound insulation of building elements*, Geneva, 1995.
2. ASTM E90: *Standard test method for laboratory measurement of airborne sound transmission loss of building partitions and elements*, 2009.
3. ISO 10140-3: *Acoustics—Laboratory measurement of sound insulation of building elements—Measurement of impact sound insulation*, Geneva, 2010.
4. M. Asselineau, Quelques aspects de l'affaiblissement acoustique d'un élément de façade soumis au bruit de la circulation (A few aspects of the sound reduction index of a facade sample exposed to traffic noise, in French), doctoral thesis, Le Mans, 1987.
5. ISO 140-4: *Acoustics—Measurement of sound insulation in buildings and of building elements—Part 4: Field measurements of airborne sound insulation between rooms*, Geneva, 1998.
6. ASTM 336: *Standard test method for measurement of airborne sound attenuation between rooms in buildings*, 2011.
7. *Acoustique—Vérification de la qualité Acoustique des bâtiments (Acoustics—Verification of the acoustic quality of buildings*, in French), Daint Denis, 1982.
8. ISO 16283-1: *Acoustics—Field measurement of sound insulation in buildings and of building elements*, Geneva, 2013.
9. ISO 140: *Acoustics—Measurement of sound insulation in buildings and of building elements—Field measurements of airborne sound insulation of facade elements and facades*, Geneva, 1998.
10. ASTM E413: *Classification for rating sound insulation*, 2010.
11. ISO 717-1: *Acoustics—Rating of sound insulation in buildings and of building elements—Part 1: Airborne sound insulation*, Geneva, 2013.
12. ISO Pr16717-1: *Acoustics—Evaluation of sound insulation spectra by single numbers—Part 1: Airborne sound insulation*, Geneva, expected 2014.
13. TNO, LGILAB computer program, 1984.
14. Gamba, Acou-stif computer program, www.gamba-logicielacoustique.fr/logiciel-acous-stiff.
15. CSTB, Acoubat computer program, www.cstb.fr/dae/fr/nos-produits/logiciels/acoubat-sound.html.
16. EN 12354: *Building acoustics. Estimation of acoustic performance in buildings from the performance of elements. Airborne sound insulation between rooms*, 2000.
17. C. Guigou, M. Villot, Prediction methods adapted to wood frame lightweight constructions, *Building Acoustics*, 13(3), 2006.

18. W.M. Schuller et al., *Contrôle du bruit en milieu industriel* (*Industrial noise control*, in French), Dunod, 1981.

19. B. Rasmussen, J.H. Rindel, Concepts for evaluation of sound insulation of dwellings—From chaos to consensus? presented at Forum Acusticum, Budapest, Hungary, 2005, Paper 7820.

20. B. Rasmussen, Sound insulation between dwellings—Requirements in building regulations in Europe, *Applied Acoustics*, 71, 373–385 (2010).

21. A. Carvalho, J. Amorim, Acoustic regulations in European Union Countries, presented at Conference in Building Acoustics "Acoustic Performance of Medium Rise Timber Buildings" Proceedings, Dublin, 1998.

22. V. Desarnaulds, B. Rasmussen, Harmonisation des réglementations européennes dans le domaine de l'isolation acoustique dans le bâtiment (COST TU0901) (Harmonization of European regulations on sound insulation in buildings, in French), presented at Proceedings of the 10th French Congress of Acoustics, Lyon, 2010.

23. Décret n° 2006-892 du 19 juillet 2006 relatif aux prescriptions de sécurité et de santé applicables en cas d'exposition des travailleurs aux risques dus au bruit et modifiant le code du travail (Decree pertaining to safety prescriptions against noise risks), *Journal Officiel de la République Française*, 166, 10905 (2006).

24. Décret n°98-1143 du 15 décembre 1998 relatif aux prescriptions applicables aux établissements ou locaux recevant du public et diffusant à titre habituel de la musique amplifiée, à l'exclusion des salles dont l'activité est réservée à l'enseignement de la musique et de la danse (French decree about prescriptions applicable to public receiving facilities and rooms using amplified music with the exception of dance and music teaching reserved rooms), *Journal Officiel de la République Française*, December 16, 1998.

25. Arrêté du 30 juin 1999 relatif aux caractéristiques acoustiques des bâtiments d'habitation (Arrest about acoustic characteristics of dwellings), *Journal Officiel de la République Française*, 149, 10658–10660 (1999).

26. Arrêté du 25 avril 2003 relatif à la limitation du bruit dans les établissements de santé (Arrest about noise control in healthcare facilities), *Journal Officiel de la République Française*, 123, 9104 (2003).

27. J. Evans, C. Himmel, Acoustical standards and criteria documentation of sustainability in hospital design and construction, presented at Acoustics 2012 Proceedings, Nantes, 2012.

28. J. Evans, Acoustical standards for classroom design—Comparison of international standards and low frequency criteria, presented at Low Frequency 2004 Proceedings, Maastricht, 2004.

29. Arrêté du 25 avril 2003 relatif à la limitation du bruit dans les établissements d'enseignement (Arrest about noise control in educational facilities), *Journal Officiel de la République Française*, 123, 9102 (2003).

30. HM Government, *The building regulations 2010—Resistance to the passage of sound, approved document E*, NBS, London, 2010.

31. *Building Code of Australia 2011*, Australian Building Codes Board, Canberra, Australia.

32. DS 490: *Lydklassifikation af boliger* (*Sound classification of dwellings*), Denmark, 2007.

33. SFS 5907: *Rakennusten akustinen luokitus* (*Acoustic classification of spaces in buildings*), Finland, July 2005.

34. IST 45: *Acoustics—Classification of dwellings*, Iceland, 2003.

35. NS 8175: *Lydforhold i bygninger, Lydklassifisering av ulike bygningstyper* (*Sound conditions in buildings—Sound classes for various types of buildings*), Norway, 2008.

36. SS 25267: *Byggakustik—Ljudklassning av utrymmen i byggnader—Bostäder* (*Acoustics—Sound classification of spaces in buildings—Dwellings*), Sweden, 2004.

37. STR 2.01.07: *Dėl statybos techninio reglamento str 2.01.07:2003, pastatu vidaus ir isores aplinkos apsauga nuo triuksmo* (*Lithuanian building regulations, protection against noise in buildings*), Patvirtinimo, Lithuania, 2003.

38. NEN 1070: *Geluidwering in gebouwen—Specificatie en beoordeling van de kwaliteit* (*Noise control in buildings—Specification and rating of quality*), The Netherlands, 1999.

39. *Code de construction du Québec, Chapitre I—Bâtiment, et Code national du bâtiment*, Canada, 2005 (modifié).

40. ISO 140-6: *Acoustics—Measurement of sound insulation in buildings and of building elements—Part 6: Laboratory measurements of impact sound insulation of floors*, Geneva, 1998.

41. ASTM 492: *Standard Test Method for Laboratory Measurement of Impact Sound Transmission through Floor-Ceiling Assemblies Using the Tapping Machine*, 2009.

42. ISO 10140-3: *Acoustics—Laboratory measurement of sound insulation of building elements—Part 3: Measurement of impact sound insulation*, Geneva, 2010.

43. ISO 140-7: *Acoustics—Measurement of sound insulation in buildings and of building elements—Part 7: Field measurements of impact sound insulation of floors*, Geneva, 1998.

44. ASTM 1007: *Standard test method for field measurement of tapping machine impact sound transmission through floor-ceiling assemblies and associated support structures*, 2013.

45. ISO 15712-2: *Building acoustics—Estimation of acoustic performance of buildings from the performance of elements—Part 2: Impact sound insulation between rooms*, Geneva, 2005.

46. JIS A 1419-2: *Acoustics—Rating of sound insulation in buildings and of building elements—Part 2: Floor impact sound insulation (in Japanese)*, Tokyo 2000.

47. N. Tselios et al., Correlation of impact isolation test results using standard tapping machine and heavy/soft impact source, presented at ICSV14 Proceedings, Cairns, 2007.

48. A.C.C Warnock, Floor research at NRC Canada, presented at Proceedings of Conference in Building Acoustics "Acoustic Performance of Medium-Rise Timber Buildings," Dublin, Ireland, December 1998.

49. DIS ISO 16283-2: *Acoustics—Field measurement of sound insulation in buildings and of building elements—Part 2: Impact sound insulation*, Geneva, 2013.

50. ISO 140-18: *Acoustics—Measurement of sound insulation in buildings and of building elements—Part 18: Laboratory measurement of sound generated by rainfall on building elements*, Geneva, 2006.

51. M.R. Connelly, Acoustical characteristics of vegetated roofs: Contributions to the ecological performance of buildings and the urban soundscape, Doctoral thesis, University of British Columbia, Vancouver, 2011.

52. ASTM E989: *Standard classification for determination of impact insulation class (IIC)*, 2012.

53. ISO 717-2: *Acoustics—Rating of sound insulation in buildings and of building elements—Part 2: Impact sound insulation*, Geneva, 2013.

54. ISO Pr16717-2: *Acoustics—Evaluation of sound insulation spectra by single numbers—Part 2: Impact sound insulation*, Geneva, expected 2014.

55. ISO 140-8: *Acoustics—Measurement of sound insulation in buildings and of building elements—Part 8: Laboratory measurements of the reduction of transmitted impact noise by floor coverings on a heavyweight standard floor*, Geneva, 1997.

56. ASTM E2179: *Standard test method for laboratory measurement of the effectiveness of floor coverings in reducing impact sound transmission through concrete floors*, 2009.

57. EN 16205: *Laboratory measurement of walking noise on floors*, 2013.

58. V. Desarnaulds et al., Swiss pendulous hammer for decoupling measurement of service equipment in wooden multi-storey building, presented at Proceedings of EuroNoise 2012, Prague, 2012.

59. SIA 181: 2006, *Protection contre le bruit dans le bâtiment (Protection against noise in buildings, in French)*, Swiss Society of Engineers and Architects (SIA), Muttenz, 2006.

60. ISO 354: *Acoustics—Measurement of sound absorption in a reverberation room*, Geneva, 2003.

61. ASTM C423: *Standard Test Method for Sound Absorption and Sound Absorption Coefficients by the Reverberation Room Method*, 2009.

62. ISO 10534-1: *Acoustics—Determination of sound absorption coefficient and impedance in impedance tubes—Part 1: Method using standing wave ratio*, Geneva, 1996.

63. ISO 10534-2: *Acoustics—Determination of sound absorption coefficient and impedance in impedance tubes—Part 2: Transfer-function method*, Geneva, 1998.

64. ASTM C384: *Standard test method for impedance and absorption of acoustical materials by impedance tube method*, 2011.

65. K.V. Horoshenkov, F.-X. Bécot, L. Jaouen, Acoustics of porous materials: Recent advances relating to modelling, characterization and new materials, *Acta Acustica United with Acustica*, 96(2), (2010).

66. EN/ISO 14257: *Acoustics—Measurement and parametric description of spatial sound distribution curves in workrooms for evaluation of their acoustical performance*, Geneva, 2001.

67. Arrêté du 30 août 1990 pris pour l'application de l'article R. 235-11 du code du travail et relatif à la correction acoustique des locaux de travail (Arrest pertaining to the acoustic treatment of work spaces, in French), *Journal Officiel de la République Française*, 244, (1990).

68. ISO 3382-3: *Acoustics—Measurement of room acoustic parameters—Open plan offices*, Geneva, 2012.

69. Arrêté du 1er août 2006 fixant les dispositions prises pour l'application des articles R. 111-19 à R. 111-19-3 et R. 111-19-6 du code de la construction et de l'habitation relatives à l'accessibilité aux personnes handicapées des établissements recevant du public et des installations ouvertes au public lors de leur construction ou de leur création (Arrest dated August 1, 2006, defining prescriptions for accessibility by reduced mobility people to public receiving facilities, in French), *Journal Officiel de la République Française*, August 2006.

70. ISO 2631-2: *Mechanical vibration and shock—Evaluation of human exposure to whole-body vibration—Part 2: Vibration in buildings (1 Hz to 80 Hz)*, Geneva, 1996.

71. M. Serra, M. Asselineau, Solid borne noise in buildings—How does one cope? presented at Joint French-English Acoustics Congress SFA IOA, Nantes, 2012.

72. OFEFP (Swiss Federal Office for Environment Forestry and Landscape Protection), *Directive pour l'Evaluation des Vibrations et du Bruit solidien des installations de transport sur rail* (*Directive for the assessment of solid borne noise from rail transportation installations*), Office Fédéral de l'Environnement, des Forêts et du Paysage, Bern, Switzerland, December 20, 1999.

73. CDM, Dephi sleeper, Commercial brochure.

74. S. Ahmad et al., Simplified design for vibration screening by opened and in-filled trenches, *Journal of Geotechnical Engineering*, 117(1), (1991).

75. M. Asselineau, M. Vercammen, Détermination du facteur de rayonnement des équipements (Determination of the noise radiation factor of equipment, in French), presented at Proceedings of the Congress Prévision du Bruit Émis par les Structures Mécaniques Vibrantes (Prediction of Noise Emitted by Vibrating Structures, in French), CETIM, Senlis, 1991.

76. http://www.dailymotion.com/video/x2q3pb_catastrophe-effondrement-du-pont-ta_fun#. UbmC7PlM-hU.

77. I. Prade, Etude pour une meilleure gestion des nuisances sonores et vibratoires dues à l'activité de chantiers de construction de bâtiments (Study for better noise control of construction activities, in French), Engineer's thesis, CNAM Paris, 2013.

78. JBS Environmental, Construction noise and vibration management plan, Project 7307, 2010.

79. Guidelines on noise control for construction sites, Health and Social Services of the States of Jersey, Saint Helier, 2004.

80. Construction requirements and guidelines, City of Toronto, 2004.

81. M. Asselineau, The challenge of heavy rehabilitation projects—case studies, presented at ICSV13 Proceedings, Vienna, 2006.

82. Placo Saint Gobain, *Doublages isolants* (Insulating doublings, in French), Suresnes, 2011.

Chapter 4

Mechanical Equipment

4.1 INTRODUCTION

Mechanical equipment nowadays is a regular part of a building. One needs to be aware of the possible noise and vibrations that may be transmitted to the living quarters of a building, or even radiated in the outside environment.

This chapter will review a few usual types of mechanical equipment as well as comment and advise on their acoustical aspects.

4.2 APPLICABLE STANDARDS AND REGULATIONS

Mechanical equipment can be covered by several law texts:

- Regulations pertaining to community noise control or to equipment classified for the protection of the environment as concerns noise radiated in the outside environment
- Regulations pertaining to the noise exposure level of workers as concerns the maintenance workers
- Regulations and standards pertaining to the work conditions in the rooms nearby
- Regulations pertaining to the noise levels generated by mechanical services inside dwellings and public receiving spaces

Usually those regulations will distinguish between permanently operating equipment (e.g., a heating boiler) and sporadically operated (e.g., the flushing system of the toilets); for example, in the French regulations [1] for the acoustics of dwellings there is a requirement of 30 dB(A) for permanently operated equipment and 35 dB(A) for sporadically operated equipment. In Switzerland, there also is a requirement for impulsive sounds generated by equipment [2]. In addition, some standards require minimal acoustic performances for the walls or floors on which equipment is mounted [3] (cf. Section 3.4.10).

4.3 HVAC

HVAC stands for heating, ventilating, and air conditioning. Basically, a HVAC system will feature a cooling or heating system for the fluid, a circulator to help that fluid circulate into the building, and terminals in the spaces of interest.

4.3.1 Boilers

Boilers produce hot water or steam using a heat source. While small individual boilers are either electric or gas burning, larger boilers burn gas, fuel, or coal.

The noise can come from the pump bringing the fuel or from the coal dropping on the chute. The burner is the main contributor; in many cases, it is supplemented by a fan to increase combustion.

The overall sound power level of a boiler is given in [4] for either a small or a large boiler as follows:

$$\text{Small boiler: } L_w = 95 + 4 \log(M/15)$$

where M is the quantity of steam produced, in kilograms.

$$\text{Large boiler: } L_w = 84 + 15 \log(P)$$

where P is the power output, in MW.

4.3.2 AHU and Fans

Fans are used to either deliver or extract air from a space. The simplest to be found is the axial fan, in which the airflow is perpendicular to the wheel. Unfortunately, while it is pretty simple to implement (we have all seen such fans in a wall or even in a window of a workshop or kitchen), it does not accept a large pressure drop. In order to cope with pressure drops, one uses a centrifugal fan in which the airflow is tangent to the wheel.

The sound power level of a fan per octave band is given by the formula [4]

$$L_w = 40 + 10 \log(Q) + 20 \log(p) - Cor$$

where Q is the flow rate, in m^3/s, p is the pressure increment, in Pa, and Cor is a frequency correction in the range 7 to 26.

More detailed models (i.e., regarding sound power levels per octave band, injected in the ducts or radiated in the fan's environment, according to type of fan unit) are available in [5, 6].

AHU stands for an air handling unit. Such units are used to circulate air inside a building, which is eventually electrically heated or water cooled prior to its circulation. Manufacturers normally provide at least the A-weighted sound power level of the ducted equipment plus the A-weighted sound power level radiated by the unit in its environment under a free field over reflective plane conditions.

4.3.3 Compressors

Compressors can be found as stand-alone equipment or as part of larger equipment.

For a rotary or reciprocating compressor, the overall sound power level radiated by the casing is given by the formula [4]

$$L_w = 90 + 10 \log(P)$$

where P is the power, in kW.

For a centrifugal compressor, the overall sound power level radiated by the casing is given by the formula [4]

$$L_w = 79 + 10 \log(P)$$

More detailed models (i.e., regarding sound power levels per octave band, injected in the ducts or radiated in the compressor's environment, according to type of compressor unit) are available in [6].

4.3.4 Heat Exchangers

Heat exchangers are equipment where heat is exchanged from one fluid to another. In some cities there are some energy production centers that will distribute steam or iced water to the plant rooms of their clients. In such a plant room the heat exchanger will then transfer thermal energy into the network of the client.

On the basis of experience, the noise contribution of such heat exchangers is negligible compared to the contribution of pumps and fans.

4.3.5 Freezers

Freezers are meant to produce cold. As such, they feature a compressor that makes noise and generates vibrations, as well as an evaporator that features a bank of fans.

Manufacturers normally provide at least the A-weighted sound power level radiated by either the combined unit or the separate units plus the A-weighted sound level radiated by the unit in its environment under a free field over reflective plane conditions. Should no such indication be available, it is possible to perform a first assessment using the models of fans and compressors from Sections 4.3.2 and 4.3.3.

4.3.6 Pumps

Pumps are typically used as circulators for the hot water acting as heat fluid in the radiators. While the noise levels generated by pumps are not as high as those of a freezer, they can generate a significant amount of vibration that will ultimately be radiated by the attached piping and the structures of the building, that is, unless proper precautions are taken. Those precautions will typically include:

- A thick (25 cm minimum) lower concrete floor in the plant room
- An inertia mass on resilient pads to rest the pump on (as illustrated in Figure 4.1)
- Flexible (e.g., Dilatoflex) links to the piping

One must beware of modern pumps that will typically feature variable speed. The lowest speed must be considered when designing the vibration control measures.

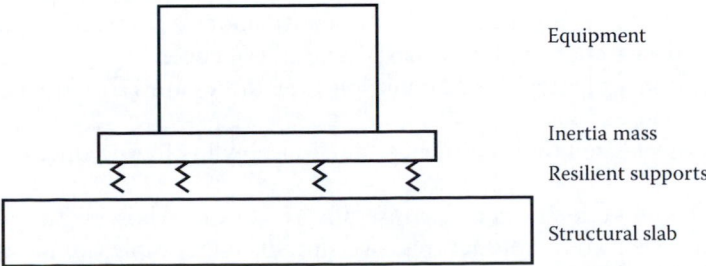

Figure 4.1 Principle of the resilient mounting of an equipment.

4.3.7 Piping and Ducting

According to their size, piping and ducting may actually radiate noise. They may also transmit vibrations. More to the point, when crossing walls and floors there can be a weakness should the corresponding hole be not properly sealed.

It is possible to estimate the sound power level radiated by a duct or a pipe from the formulas given in [6, 7].

It is possible to have the ducts and pipes located inside an enclosure (e.g., plasterboard layers with mineral wool). One should note that such enclosures will have to be designed in cooperation between the acoustic engineer and the safety engineer, as they may be of influence for fire control purposes.

Some manufacturers have absorptive ducts in their catalog: those are typically made of high-density mineral wool with an aluminum foil on the outside and a glass cloth on the inner side. While such ducts are quite light and easy to modify locally, they also feature poor wall insulation, especially in the low-frequency range, and can either radiate or receive sound energy.

4.3.8 Silencers

In order to reduce the noise energy transferred through the ducting network there are some specific noise control elements named silencers or attenuators that are inserted in the network, typically close to the noisy element (e.g., AHU). The crudest silencer simply is a piece of duct that has been internally lined. One can also use a core absorptive material inserted in the duct. There also are rectangular silencers with absorptive baffles. The longer the silencer, the more attenuation it will give. Also, the narrower the air channels, the more attenuation the silencer will provide.

One usually uses standard ISO 7235 [8] to perform a serious sound insertion loss measurement of a silencer for a given flow rate. However, due to the complication of the test rig (which typically needs at least 21 m to accommodate the fan), its silencer and the sound source plus test duct manufacturers will frequently give a static attenuation for their equipment (i.e., an attenuation measured without airflow, e.g., according to ISO 11691 [9]). One should remember that the airflow will generate turbulences (noise). More to the point, due to the reduction of airflow channel section due to the absorptive material, there is a need to increase the ducting section at the silencer (a typical industrial silencer will feature 200 mm thick absorptive baffles with 100 mm air channels). This means that a divergent and later a convergent piece of duct will be needed to connect the silencer with the ducting network. Tests *in situ* are harder to perform, but guidelines are given in standard ISO 11820 [10].

To be effective there are quite a few rules to be followed:

- Do not exceed a certain speed limit in the air channels (typically 10 m/s) in order to prevent the occurrence of noise generation in the silencer.
- Do not position a silencer at a distance less than three times the larger cross-dimension of the duct.
- Beware of short-circuiting the silencer due to feebly insulating duct walls.

A crude way to make a silencer is to use absorptive materials on the inner side of the duct walls. There even are manufacturers who produce a complex made of an aluminum foil on the external side and a glass cloth on the internal side with a rigid mineral wool in between [11], while other manufacturers produce flexible ducts along the same principles.

One should remember that while such ducts do attenuate, they can also radiate or pick up noise in the spaces they go through.

4.4 ELECTRICAL

4.4.1 Transformers

Transformers are used to reduce the voltage from the supplier to a value usable inside the building (typically 220 V in Europe and 110 V in North America). Due to constriction, they generate tonal noise (typically 100 Hz in Europe and 120 Hz in North America, together with a fourth harmonic).

Due to their heavy weight and their being at the interface between the supplier and the user, they usually are located at the ground floor. They are normally mounted on resilient rollers supplied by either the manufacturer or the energy provider. In order for those vibration control measures to be effective, the structure must be stiff enough (e.g., reinforced concrete 25 cm thick).

Care must be exercised regarding the sound generated by the equipment: To start with, the natural ventilation of the plant room leads to rather high sound levels generated outside, and one must make sure that they will be compatible with the immediate environment. Next, one may care to remember that the sound reduction index value in the 125 Hz octave band is lower than the single number rating value; it is better to make sure that the sound pressure level transmitted to the neighboring spaces is compatible with the targeted values (especially if compliance with a frequency contour such as NR or NC has been specified).

A Word of Caution: Under any circumstance *do not* enter a transformer technical room in service unless you have the proper electrical qualifications for it.

The A-weighted sound level at 1 m distance generated by a transformer can be estimated using the following formula [12]:

Dry transformer: $L_p = 39 + 9 \log(P)$

Oil transformer: $L_p = 31 + 8 \log(P)$

where P is the power in kVA.

4.4.2 Generators

Generators are used to provide electricity to a building. While their use is often limited to emergencies (e.g., operating the minimum lighting of the building and eventually one of the lifts), they are sometimes also contracted to provide electric energy to the main supplier. This will actually be one of the first questions to ask to the end user: Is the generator merely used for emergency lighting in order to give the occupants a chance to get out (which means that it will not operate for long, and at any rate, there probably will be other sources of noise, such as firemen's sirens), or can it be used for hours on end to operate some equipment? According to the answer, the noise reduction measures will most probably not be the same.

There are many types of generators, from the small transportable unit fitting in a car to the monster as large as a ship engine. Accordingly, the noise levels generated can range from 65 to 115 dB(A) at 3 m.

Should the equipment be located inside the building, one will have to consider a choice between a generator under enclosure and a standard generator. The enclosure can bring an attenuation of up to 15 dB(A), which is far from negligible should a maintenance worker be active nearby. Unfortunately, it also may complicate the life of the maintenance workers who have to dismantle part of it, and it significantly increases the overall size of the equipment. Whatever the case, a set of silencers will be needed on the ventilations of the plant room and another one on the exhaust of the motor. Access to the plant room (typically a 25 cm thick concrete structure) will be through air lock-mounted doors (for both acoustic and fire safety reasons).

4.4.3 Terminals

Terminals are the usual plug receivers, switch housings, and lighting fixtures that can be found in a space. Most of the time, such terminals will not generate noise (unless they feature a small transformer). But they may weaken the sound reduction index of the partition in which they are mounted unless proper precautions are taken (Note: this is also true of the fire resistance of the wall or ceiling of interest!). As a rule of thumb, one should not have them back-to-back (i.e., leave at least 60 cm between them). When this proves impossible, one must try to reconstitute the wall or ceiling leaf behind the terminal box. If this still is too much of a delicate work, the last chance will be to coat the external sides of the terminal box using plaster.

4.4.4 Wiring

Wiring does of course not generate noise. However, the crossing of walls and floors may constitute a weakness in the sound insulation unless proper measures are taken (i.e., filling the remaining cavity). When properly done, the wiring will typically run through a small- to medium-diameter pipe that will be included in the wall using the same type of material (i.e., plaster for a plasterboard partition or ceiling, concrete for a concrete wall or floor, etc.). Prior to being forced into a final position, the pack of wires will be wrapped using mineral wool.

4.4.5 Batteries and Static Convertors

Batteries will of course not generate noise. But they usually are located in a room where there will be an exhaust fan whose noise radiated to the outside environment (and sometimes even in the building; cf. Section 4.9.7) must be controlled. More to the point, there will often be an inverter nearby that will need some vibration control measures (i.e., mounting on a rigid steel frame resting on resilient pads).

Static convertors can be used to produce DC current out of the AC mains (the simplest of the static convertors is a fixed rectifier). In addition, they can be used to regulate the output voltage. While basically the main components are made of static elements (diodes, thyristors, etc.), there still are some elements that may generate noise and vibrations due to constriction (e.g., a winding or inductance). Therefore, it is necessary to provide vibration control measures for such an apparatus (cf. Section 4.9.9).

4.5 LIFTS

4.5.1 Various Types of Machinery

Lifts can be operated using different kind of machinery. Here are a few possibilities:

- Hydraulic lifts: They call for a hydraulic system pushing up or down the cabin. Those usually are reserved for a limited number of floors. They require a machine room usually located by the bottom of the shaft.
- Lifts with a separate machine room: The machine room typically is located over the shaft or by its upper end. This is quite typical of older installations. Apart from the motor and gearbox noise and vibrations, one may also have to worry about the noise and vibrations from the electromechanical switches earnestly clanging away.
- Lifts with onboard machinery: These have developed over the last years. While they often feature limited vibrations, the noise from the machinery located on top of the cabin traveling with it has to be accounted for.
- Lifts within shaft machinery: The machinery typically is installed inside the shaft near its top. While this is a space-saving scheme, one has to make sure of the vibration control measures as well as take relevant precautions for the safety of the workers.

One may sometimes find fancier systems, such as an inclined lift, which is more akin to a steep cog railway.

Whatever the system considered, the manufacturer should be able to provide some basic acoustic data, such as the sound level inside the cab and inside the shaft, as well as the velocity or acceleration levels on the supporting structure [13, 14].

4.5.2 Shaft

The shaft should be made of at least a 20 cm thick reinforced concrete wall. When a bedroom or living room or other sensible space is next to it, then this wall should be upgraded using a material with a sound reduction improvement index ΔR greater than 5 dB (e.g., a 13 mm plasterboard on a 50 mm factory assembled on a mineral wool).

Doors to the shaft are not to be overlooked: Unless properly treated, the slamming of such doors can prove to be a real source of nuisance. One should seek progressively shutting doors to avoid such a noise.

4.5.3 A Few Practicalities

The guiderails should be fixed at the junction of floors and walls, as this is the location with the most inertia.

A Word of Caution: While most manufacturers are able to supply basic acoustic information such as the sound power level of the machinery and the sound levels inside the shaft and inside the cab, they usually cannot provide information regarding the amount of vibratory energy injected inside a concrete structure (e.g., as per ISO 10816 or 9611 [13, 14]). This means it will probably be necessary to perform some measurements on a similar installation.

Another Word of Caution: Nowadays lifts are quite talkative, with numerous announcements forwarded near the doors (e.g., "call activated," "cab coming," "opening the doors," "going down," etc.). While this is a help for vision-impaired people, one must take such noise contribution into account inside the nearest premises.

4.6 SANITARY EQUIPMENT

Sanitary equipment is an important feature of a building. It is present at practically all floors. Inevitably, it can generate noise at all floors, and even propagate noise and vibrations throughout the building.

4.6.1 Taps

Taps may generate vibrations in the ducting system due to the turbulences inside; they will depend on the constitution of the tap as well as on the operating conditions (i.e., available pressure and flow rate).

The sound power level generated by a tap in a room can be tested under laboratory conditions according to standard ISO 3822 [15]. Manufacturers are supposed to be able to provide the acoustician with such data.

4.6.2 Pumps

Pumps are typically used to increase the available pressure at the tap. While the noise levels generated by pumps are not that high (unless no proper maintenance has been carried out, that is!), they can generate a significant amount of vibration that will ultimately be radiated by the attached piping and the structures of the building unless proper precautions are taken. Those precautions will typically include:

- A thick (25 cm minimum) lower concrete floor in the plant room housing the pumps
- An inertia mass on resilient pads to rest the pump on
- Flexible (e.g., Dilatoflex) links to the piping

4.6.3 Piping

According to their size, pipes may actually radiate noise. They may also transmit vibrations. More to the point, when crossing walls and floors, there can be a weakness should the corresponding hole be improperly sealed.

One will make a difference between supply water and wastewater systems. The former relies on high-pressure rather than small- diameter (e.g., 3 cm) ducts, while the later mainly relies on gravity and large sections.

Various materials are available for pipes. While copper (and sometimes even PVC nowadays) can be found for the supply pipes, various materials (ceramic, cast iron, or PVC) can be found for wastewater. Of course, the heavier the material, the lower the noise radiation. One may note that according to the size of the pipe, some additional specifications (e.g., for fire safety purposes or to prevent clogging) may apply.

It is possible to estimate the sound power level radiated by a pipe from the formulas given in [6, 7].

It is possible to have the pipes inside an enclosure (e.g., plasterboard layers with mineral wool). One must make sure to have trapdoors at sensitive locations (e.g., close to a bend) to enable maintenance work.

4.7 VIBRATION CONTROL

Whenever dealing with mechanical services vibration control precautions must be considered, as such vibrations may result in sound generation nearby. This usually is performed through a thick heavy structural floor and resilient supports; those supports are devised according to the lowest excitation frequency and the load.

Quite often an inertia base (also called inertia mass) is used too (cf. Figure 4.1); it has the advantage of applying better force on the resilient supports. More to the point, it will be a help when the equipment starts deteriorating over the years. Now for the bad news: It can be quite heavy, and it does take space too, so this has to be specified right at the beginning of the project. Typical pieces of equipment to be treated are pumps and freezers.

Whatever the solution used, one may care to note that several building regulations require the possibility for workers to visit and repair the waterproofing layer on the floor without undue complications. This typically results in an 80 cm clearance between the bottom of the equipment (or its inertia base) and the top of the floor. More to the point, there often is some thermal insulation under the waterproofing on the floor, and it will probably be less stiff than the resilient supports. This means that specific details must be discussed with the HVAC engineer and the structural engineer.

Last, do remember that in order to decouple the equipment, one also has to provide resilient couplings on the various ducts and cables that link it to the various mechanical networks.

4.8 A FEW PRACTICAL ASPECTS

As a general rule, the following physical points should be kept in mind when dealing with mechanical equipment:

- Apply lubrication (a badly lubricated mechanism will not only squeal, but it will also wear out more quickly).
- Most of the time speed means noise (slower equipment will usually be quieter than faster equipment): Try to opt for equipment with reduced speed.
- Provide adequate ventilation to the equipment (A badly ventilated equipment will not only have whining fans but also overheat).
- Make sure the section of the ventilation ducts is large enough (if it is not, the speed will be too high and there will be noise generation).
- Provide resilient supports under the equipment to keep vibrations from spreading around.
- Provide acoustic absorption close to the equipment (just imagine the acoustic energy radiated from the equipment and try to intercept it).
- Make sure there is easy access for maintenance purposes (badly maintained equipment will be noisier than properly maintained equipment); more to the point, if there are noise control elements such as an enclosure to be dismantled prior to maintenance, you can bet it will not be properly reassembled later on).
- Try to find a location for the equipment that is close enough.

4.9 EXAMPLES

4.9.1 Coal Boiler

As befitting its status, an old building belonging to a miner's association in Paris was using a coal-fired boiler located in the basement of the building. However, the new tenants quickly complained about the noise of the installation. An acoustician was called. It was found that the noise originated from the draught fan, but also from the automatic discharge of coal. The problem was solved through the use of silencers on the fan and a better spreading of the coal discharge over time.

4.9.2 Gas Boiler

A newly installed gas boiler was the subject of complaints by the neighborhood. It was found that the noise originated on the one hand from the draught fan, and on the other hand from the resonance of the chimney, with the latter emitting a tonal noise. The problem was solved by treating the former with regular silencers, while the latter was fitted with a quarter-wave silencer.

4.9.3 Silencer

In an industrial plant an acoustician was called to try to solve a problem of noise generated in the nearby facilities by the ventilation of a building.

The intervention was quickly performed and the report simply read: suppression of the silencer! As it happened, the silencer was causing too sharp a pressure drop—hence the amount of noise generated.

4.9.4 Ducting

On attempting to commission an office building the acoustician was startled to be treated to a siren-like noise coming from one of the rooms. On checking the origin of that 70 dB(A) noise, it was found that the contractor had adapted two pieces of duct into a junction located right after a sharp bend, which was behaving like a flute instrument.

4.9.5 AHU Noise

During the fitting out of a small bar a neighbor politely visited the building site and remarked that something had to be done about the noise from the new air handling unit he could see from his window, whose noise nearly prevented him from sleeping the previous night. After having him repeat the remark, a somewhat grinning contractor pointed out that the electric power cable was not yet connected! The bewildered neighbor took the point and apologized, but also remarked that he definitely was hearing a new ventilation noise. The acoustician accompanied him to his flat and discovered that while there was a ventilation noise, it actually came from a restaurant exhaust that had been in operation for ages.

Lesson Learned: The moment a piece of equipment can be seen, the neighbors will feel they hear it too.

4.9.6 HVAC Noise

The manager of a recently opened European luxury hotel was worried when a few clients came forward and expressed their concerns regarding the noise from the HVAC in their room. However, it quickly turned out that they were used to noisy operations in their Arabian, Asian, or American usual resorts, and they initially had trouble believing the HVAC was properly running, even when the temperature inside was 18°C while outside it was 35°C!

Lesson Learned: Beware of cultural aspects when setting the acoustic noise level limits of the mechanical equipment.

4.9.7 Fans

A large public building dating back to the end of the 19th century had steadily been upgraded over the years. One of those early upgrades had seen the creation of a battery room that needed a small exhaust fan. After the completion of recent rehabilitation work to enhance the acoustics of the various rooms, everybody was puzzled to hear a small but steady purring noise all over the place.

It eventually turned out that due to the better sound insulation, the background noise levels were lower than before, and that small noise actually came from the small exhaust fan of the battery room, which was generating vibrations in the structure!

Lesson Learned: Size does not always matter when it comes to generating noise and vibrations.

In that same building a plant room housing a fan had to be built over an office space. The structural engineer was especially unhappy, as the acoustician had already required a 25 cm slab further in the building for a freezer plant room. Calculations showed that the fan room floor could be built using a 20 cm thick concrete slab. During the building process the acoustician had checked that the contractor had properly selected the fan and got the correct silencers and resilient supports for his fan; however, due to planning difficulties, the acoustician and the head contractor were not able to inspect the whole assembly, especially as the fan was now hidden from the entrance of the plant room by a large duct transiting through that space. To complicate matters, it had been freshly sprayed, so going any further in the room was postponed.

When a 50 dB(A) noise appeared in the office due to the operation of the fan everybody came to suspect that the acoustician's calculations were wrong, as all precautions had been taken. Using a sound source, it was demonstrated that a 90 dB(A) sound generated in the fan room would not produce more than 35 dB(A) in the office underneath. More to the point, the sound level in the 125 Hz octave band differed only by 1 dB in the office and the fan room during fan operation. This had to be a structure-borne transmission. The contractor and the acoustician went crawling under the large duct. Good news: The resilient supports were installed. Now for the bad news: They were hanging from each corner of the fan's frame, which was resting on four bricks. It turned out that the workers had used the bricks to set the fan's frame at the correct position for installation and then forgot about them.

Lesson Learned: Never accept a work without visually (and acoustically!) checking it.

4.9.8 Terminals in a Wall

While the recommended practice is to keep terminals of differently owned spaces distant by more than 60 cm, it is not always considered feasible by builders or operators in such building projects as hotels and hospitals where there usually is a lot of symmetry. This will induce a weakness in the sound insulation performances of the premises.

As an example, in a luxury hotel project there were a group of four plugs and a group of two switches back-to-back in a 5 m long, 20 cm thick separating wall between rooms. Whether the wall was made of concrete or plasterboard plates, the result was similar: Without the terminals the sound insulation value was 57 dB, while it was only 53 dB with the terminals.

4.9.9 Static Convertor in a Cinema

A recently built cinema featured a small electrical room with a switch panel distributing electricity throughout the facility. In addition, there was a static convertor providing DC current to some equipment.

When the commissioning process started everybody was surprised to perceive rather loudly the noise from that small piece of equipment in the projection theatre above it (that featured a 28 cm concrete floor), that is, until the electrician realized that the resilient supports had not yet been installed!

Lesson Learned: Vibration control precautions are important for noise control even when there is significant concrete thickness.

4.9.10 Transformers Generating Noise Upstairs

In another recently built cinema there were two transformer rooms under the office area, for the cinema and the township, respectively. Shortly after the start of the commissioning procedures the occupants of the office complained about an unpleasant electric noise. When the acoustician investigated, he found that there were strong stationary waves inside, with an antinode located in the middle of the room! It eventually turned out that the township had not cared about the required resilient supporting of its transformer, which sat rigidly on the floor.

Lesson Learned: Do not trust other parties to do the required vibration and noise control work.

4.9.11 Lift

In a European building, an old lift dating back to the 1940s was upgraded. The upgrade concerned the motor controls (which had relied on electromechanical devices and were from now on operated by static heavy electronic devices). They also concerned the guiding system of the lift cabin (which previously featured leather-covered shoes sliding on steel guides).

On completion, the contractor was surprised to hear the inhabitants complaining about the noise. It was found that previously the complaints were mainly confined at the last floor dwelling, where the owner was subjected to the clanking noise of the electromechanical controlling system. With the new variable-speed controlling system, there was some electric whining sound, as the motor did not always like the DC current provided by the system, and low-frequency rumble, as the motor speed would match the frequency of the lift machinery room floor slab. This changed his soundscape for sure. In addition, dwellers at the other floors were now treated to the noise of the Teflon guides knocking on the asperities of the rail guides (e.g., junction between rails); previously, the leather guides could easily take care of such singularities. To solve the problem, a stiffening of the structure of the rail guides as well as careful grinding was required, while the floor of the lift machinery room had to be reinforced (which fortunately proved possible), together with a change toward more suitable resilient supports capable of filtering the lowest excitation frequency.

Lesson Learned: Modernizing is not necessarily synonymous with more silent.

REFERENCES

1. Arrêté du 30 juin 1999 relatif aux caractéristiques acoustiques des bâtiments d'habitation (Arrest dated June 30, 1999 pertaining to the acoustic characteristics of dwellings), *Journal Officiel de la République Française*, 163 (1999).
2. Ordonnance sur la protection contre le bruit du 15 décembre 1986 (Ordinance on noise control dated December 15, 1986, in French), Geneva, 1986.
3. SIA 181:2006: *Protection contre le bruit dans le bâtiment (Protection against noise in buildings*, in French), Swiss Society of Engineers and Architects (SIA), Muttenz, 2006.

4. I. Ver, L. Beranek, *Noise control*, J. Wiley & Sons, New York, 1992.

5. Peutz, *Eole* (Program for predictive computation of noise levels from ventilation systems), Internal report, 2004.

6. Peutz, *Noise from gas handling systems—Predictive computational model*, Internal report, 1996.

7. CONCAWE, *The prediction of noise radiated from pipe systems—An engineering procedure for plant design*, The Hague, 1987.

8. ISO 7235: *Acoustics—Laboratory measurement procedures for ducted silencers and air-terminal units—Insertion loss, flow noise and total pressure loss*, Geneva, 2003.

9. ISO 11691: *Acoustics—Measurement of insertion loss of ducted silencers without flow—Laboratory survey method*, Geneva, 1995.

10. ISO 11820: *Acoustics—Measurement on silencers in situ*, Geneva, 1996.

11. Isover, Climaver, Commercial brochure, Paris, 2002.

12. W.M. Schuller et al., *Contrôle du bruit en milieu industriel* (*Industrial noise control*, in French), Dunod Paris, 1981.

13. ISO 9611: *Characterisation of sources of structure-borne sound radiation from connected structures—Measurement of velocity at the contact points of machinery when resiliently mounted*, Geneva, 1996.

14. ISO 10816: *Mechanical vibration—Evaluation of machine vibration by measurements on non-rotating parts*, Geneva, 2009.

15. ISO 3822-1: *Acoustics—Laboratory tests on noise emission from appliances and equipment used in water supply installations—Part 1: Method of measurement*, Geneva, 1999.

Chapter 5

User's Equipment

5.1 FOREWORD

The acoustician usually does his best to ensure that the building he is studying is of proper acoustic quality, as may actually be required by the applicable sustainable development project standards chosen by the end user. However, this is not sufficient to be assured of low noise level values due to the multiplicity (and sometimes poor quality too!) of the user's equipment later brought inside the premises.

This chapter will review a few usual types of user's equipment as well as comment and advise about their acoustical aspects.

5.2 APPLICABLE STANDARDS AND REGULATIONS

According to the actual situation (location and use) of the particular piece of equipment, some standards and regulations, as well as specific contractual requirements, may apply:

- Community noise control texts regarding the noise emitted by appliances and radiated or transmitted to the neighbor's property (either indoors or outdoors)
- Labor and health regulations pertaining to the noise emission of equipment or appliances
- Contractual requirements regarding the noise emitted by a given piece of equipment

Those texts will usually result in a limitation on the sound power level or on noise levels emitted at a given location. They may even introduce a restriction on the schedule of operation.

5.2.1 Occupational Regulations

The basic idea is to prevent the equipment from generating noise levels that may be harmful to the user (or to people in the vicinity).

EU Directive 89-391[1] and later on EU Directive 2006/42/EC [2] requires equipment to display the value of the noise level at the operator's position when it is greater than 70 dB(A) as measured under the relevant applicable measurement standard. In addition, when this value is greater than 80 dB(A), it must display the sound power level as measured under the relevant applicable measurement standard.

5.2.2 Community Noise Control Regulations

Typical community noise control regulations will typically state a schedule of operation for noisy operations (e.g., forbidding operations before 10:00 a.m. and after 4:00 p.m. on weekends, as well as before 8:00 a.m. and after 7:00 p.m. on weekdays).

In addition, there will be a noise limit (typically to be assessed according to the type of area and the period of the day; the early version of standard ISO 1996 provided guidance on the sound level values and gave examples of such values [3] to be complied with at the edge of the property). This is no longer the case, as this is the task of regulations (that may rely on the standard for assessment and verification purposes though); as befits a standard, ISO 1996-1 does not provide examples of such values [4].

Some countries also use a quantity that describes the excess of particular noise over the background noise level. This quantity is defined as the difference between the ambient noise level when the equipment is operating and the background noise level measured when it is not.

It is called emergence in the recent (2014) FDIS ISO 1996 [5] (please refer to Section 2.8.5 for more information on this topic). For example, in France a recommendation by the Health Ministry states that a noise from an equipment or activity can be considered potentially annoying when its emergence value is greater than 5 dB(A) in daytime and 3 dB(A) in nighttime [6]. The local authorities can give precise information regarding those restrictions and their eventual mitigating circumstances (e.g., limited duration or emergency operation).

5.2.3 Contractual Requirements

Contractual requirements will typically cover requirements that are in excess of the legal requirements, or simply not covered by legal aspects. For example, this may concern the noise from pieces of equipment inside a performance hall or in a room not covered by regulations (e.g., director's office).

5.3 A FEW USUAL TYPES OF USER'S EQUIPMENT

This section reviews a few usual types of user's equipment and points out a few acoustical points of interest.

5.3.1 Stage Equipment

Stage equipment is usually not a source of problem for occupational or community noise regulations. However, when enjoying a show in a performance hall, one is not really keen on hearing the noise from a motorized hanger above the music or in the silence! This means contractual requirements will be issued.

With a background noise level often under the 25 dB(A) mark, this means there are serious noise control measures to be implemented. Typically, the motors will be enclosed in a dedicated technical room with the cables running out of it through a small absorptive tunnel and over a well-lubricated pulley.

It is not unusual for performers to bring in their own equipment. Usually those are motors with an acoustic enclosure around them and resilient mounting. Other pieces of equipment are often fitted with resilient hangers so as to minimize vibration transmission.

Should a user require a new piece of equipment, it is necessary to ask for a minimum of data in the call for tender documents. Such data should include at least the sound power level of the equipment together with the description of the measurement conditions.

5.3.2 Production Equipment

Production equipment typically is covered by occupational noise regulations. In addition to the usual requirements regarding the sound exposure level of personnel (cf. Chapter 10), there often are some specific regulations limiting the noise levels emitted by such appliances as grinders or lawnmowers or other.

In addition, community noise control regulations may apply too (e.g., it is ill advised to start hammering or mowing in the late evening hours).

One might remember that when delivering a piece of equipment, the manufacturer will outline under which conditions the equipment is supposed to be operated. Failure to leave on the acoustic enclosure (e.g., for ease of access to various parts of the equipment) usually is considered faulty behavior in law.

When selecting the equipment, specifications must have been issued to the tender regarding the acoustic data to be provided to the potential client (e.g., sound power level or noise level at a given distance and under specific load or operating conditions). The industry standards usually have such templates available as well as a list of the relevant standards.

5.3.3 Computer and Video Equipment

While the noise emission of computers and video equipment may seem rather low, they often represent a significant part of the soundscape of an office. A sound pressure level value of 45 dB(A) at 1 m is not uncommon for the main unit of a computer; when dealing with an open-plan office, it definitely is part of the background noise.

When it comes to video equipment, a video projector can easily generate 60 dB(A) at 1 m due to its cooling fan. This is especially a pity when good care has been applied to the design of a meeting room.

In both cases it is possible to implement reasonably good soundproofing enclosures. But one must keep in mind that a proper ventilation of the equipment must be implemented.

5.3.4 Personal HVAC Units

Personal HVAC units are quite ubiquitous nowadays, and one can often hear them purring when warm days come. In France they account for close to 40% of all community noise complaints. They come under the community noise control regulations and may also be covered by standards pertaining to the noise level inside premises. One should also remember that they can easily generate vibrations in a light structure.

Installation of such units should not be carried out lightly. First, one should assess the background noise on site, both in the outside environment and inside the premises. Next, a predictive assessment of the noise levels generated by the unit should be performed, and the proper noise control measures elaborated (e.g., a noise barrier). It is especially ill advised to try to install such a unit in direct view of the neighbors, as they will feel readily annoyed by it.

5.3.5 Electric Generators

There is a broad range of equipment falling under this label, from small portable units to very large pieces of equipment occupying half of the basement or the roof in a building.

According to their size and their use, they will have to comply with occupational noise regulations, environmental regulations, community noise control regulations, etc.

Noise control of such equipment can be carried out using the relevant soundproof enclosures and silencers. Usually the manufacturer of such equipment has such items in his catalog.

5.3.6 Sound Systems

Now this is a modern life ubiquitous piece of equipment. Unfortunately, the lawmakers who devised dwelling acoustic regulations did not necessarily think of it when stating recent regulations! More to the point, the frequency range and power output have steadily grown over the years, and this is bound to create problems.

There are a few regulations applicable on the subject: community noise control of course, and even occupational noise regulations pertaining to noise exposure. Some countries have issued specific regulations regarding public receiving spaces where amplified music is played. Such regulations usually feature a minimum sound insulation requirement to protect the neighborhood; in addition, they usually define a maximum sound level in the public accessible areas.

When it comes to a simple sound system operated by an individual community, noise regulations apply. In addition, in a multidwelling building there may be some additional requirements in the internal contractual rules in force.

5.3.7 Miscellaneous Outdoor Noise Sources

Most outdoor noise sources, from the lawn mower to the dog through the eventual sawing or hammering or music playing, are under the community noise control regulations in force. Usually the township has the relevant national or local regulations available for inquiries.

5.3.8 Selecting the Equipment

Selection of the equipment must be carried out according to several factors (e.g., pricing, reliability, ease of transportation if need be, etc.). Regarding the acoustical aspects, the following points must be investigated:

- Noise emitted in the environment. (Does it comply with the regulations in force regarding community noise control?)
- Noise generated inside the premises. (Does it comply with the regulations in force regarding community noise control? Does it satisfy the recommendations of the applicable standards regarding comfort in the premises?)
- Noise at the workstation. (Does it comply with the occupational noise regulations in force?)

In all cases, the measurement results must be accompanied by a description of the measurement conditions.

5.4 A FEW PRACTICAL ASPECTS

When dealing with the user's personal equipment, a few practicalities may be considered when dealing with noise control:

- The location of the equipment is of importance: It is particularly ill advised to install a noisy piece of equipment under the nose of the neighbor! More seriously, one should avoid installing such equipment on a light structure that could be excited by either its vibrations or its noise. On a practical side, one may care to install it as close as possible to its place of utilization to help minimize the noise of people going back and forth, yet try to keep it from noise-sensitive areas.
- Ease of access for maintenance purposes should be a factor (badly maintained equipment will be noisier and its life expectancy may be reduced too).

- Acoustic absorption close to the equipment definitely helps by eliminating potential acoustic reflections on the surfaces so treated. One should try to use an absorptive material that is shock resistant and easy to clean, as well as compliant with the relevant regulations (e.g., fire safety).
- Vibration control precautions must be considered, as such vibrations may result in sound generation nearby. This usually is performed through a thick heavy structural floor and resilient supports; those supports are devised according to the lowest excitation frequency and the load. Quite often an inertia mass is used too.
- Ventilation of the equipment is a key factor both for noise control and for the longevity of the equipment. While locating the equipment very close to a wall may seem a good thing for space planning purposes, it will probably complicate its ventilation, as the air intake at the back of the equipment will be partially clogged, and that will result in louder ventilation noise as well as overheating.
- Lubrication of equipment will reduce the efforts from the driving mechanism (thus reducing noise) and eliminate squeaking noises too.
- The implementation of an enclosure can be considered for sure. However, all factors must be considered prior to committing. Useful hints can be found in standard ISO 15667 [7]; the following points are especially worth considering:
  - Safety aspects
  - Targeted sound attenuation
  - Ease of access

5.5 EXAMPLES

5.5.1 Ceiling Fan

The acoustician was called by a property investor: One of his tenants was complaining about a strange low-frequency noise in the bedroom located at the third floor of a rather high standing building. When assessing the situation, it was found that there actually were stationary waves in it! As a temporary measure the acoustician could only suggest moving the bed to avoid being close to a large-amplitude location. Next, the hunt for the noise source was on. The tenants were quite adamant that the low-caste grocer at ground level was surely responsible due to the cooling system of his store. However, the acoustician noticed that when climbing the stairs, the incriminated noise could not be heard at ground level (next to the grocery) but was clearly perceptible at the second floor. More to the point, the grocer's equipment was observed, under his kind guidance, to be properly noise controlled. It eventually turned out that the tenants at the second floor had recently installed a ceiling fan that was not properly balanced and generated vibrations in their upper floor. The problem was solved the moment the fan was properly balanced.

Lesson Learned: Do not blindly incriminate people. Try to be logical in the diagnosis.

5.5.2 HVAC Unit in a Courtyard

The owners of the top-floor dwelling in a multipurpose building came visiting the ground level of the building where a shop was under heavy rehabilitation. They were complaining of the noise from the recently installed HVAC unit, which was visible from their window on the courtyard side. A worker had to explain to them that with no power cable yet installed, it certainly was not the noise from that unit!

On visiting their dwelling the acoustician helped them discover that the noise they heard actually came from the extractor of the restaurant next door that had been operating for years without their noticing it!

Lesson Learned: When a piece of equipment is visible, it is felt to be noisy.

5.5.3 Outdoor Heat Pump

The owner of a villa decided to have air conditioning using a heat pump. The contractor proposed to have it installed on the roof. Complaints were quick to come from every side: The urban inspectorate of the township remarked that no proper authorization had been requested for a project that changed the appearance of the roof, while neighbors complained about the noise; more to the point, the wife and children of the owner complained about the noise generated inside the premises. Facing a negative court of justice decision, the owner decided to move the heat pump to ground level close to a corner. While this move tamed the urban inspectorate, it left the neighbors irate, as there still was significant noise to be heard around. An acoustician was eventually called and prescribed the necessary absorptive noise barrier to be erected in front of the machine, together with some extra absorptive cladding on the portion of façade close to it. At last the neighbors were not complaining anymore; furthermore, after a year of operation the manufacturer had to admit that his systematic rejection of noise barriers close to the equipment was unfounded as long as proper spacing was maintained in order to enable proper ventilation.

Lesson Learned: Call a professional from the beginning, not a simple contractor.

5.5.4 Rectifier

A company needed rectifiers to produce direct current. Those were duly installed at their floor. To their surprise, the neighbor quickly remarked that he was hearing a tonal noise all over his office and asked what they intended to do about it. The problem was quickly solved through the use of a small inertia mass on resilient pads to filter the 100 Hz vibrations generating this noise (please note this was in Europe; 120 Hz should be expected in America and Asia).

5.5.5 Generator

A company decided to have an electric generator installed on its premises in order to ensure a proper emergency power supply to its batteries. A contractor was duly awarded the job of installing two 750 kVA units.

The initial project called for this equipment to be installed in the basement of the building. Fortunately, someone among the end user team spotted a few discrepancies in the physical data submitted by the manufacturer. Regarding the noise emitted by the equipment, the manufacturer explained that a generator could be delivered either with or without cladding, with corresponding noise levels in the generator room being either 110 dB(A) or 90 dB(A). The technician of the end user screamed that he did not like the clad version due to the complications involved in maintenance; more to the point, the dimensions of the clad generator were too large for comfort in the basement. An unclad version was considered until the labor inspectorate hinted they would not like it.

The equipment was eventually sent on the terrace roof of the building, which had provisions for such pieces of equipment. Each unit came as a small soundproof container with a small chimney fitted with silencers, with a sound level of 75 dB(A) at 2 m from the exhaust. This satisfied everybody involved.

5.5.6 Lubrication

The acoustician was called in a case involving a couple of innocuous-looking sheds that had been turned into a grain elevator facility. As the operator had been slow responding to the complaints, the situation was already a bit embittered due to the neighbors being exposed to noise and dust.

Various noise control measures were implemented until so-called final measurements could take place. Both the acoustician and the authorities had a bit of a laugh when the neighbors declared themselves quite satisfied, yet the sound level meter readings were a bit too high. It turned out that the handling equipment was loudly squeaking due to the lack of lubrication!

5.5.7 Piano

A good pianist took quarters in a recently built condominium. Two days after he had played in his flat he received a couple of infuriated registered letters. He answered back at once to acknowledge and promise not to play again until a proper solution was found.

A specialized contractor delivered a dedicated cabin that enabled the noise from the piano, both structure borne and radiated, not to invade the neighboring flats. No complaint was ever heard again.

Lesson Learned: Acting at once can save a lot of trouble.

5.5.8 Sound System

A good family took quarters in a recently built condominium. Unfortunately, the son loved his sound system playing loud music. Complaints came quickly through infuriated registered letters from neighbors. While the family pondered the possibility of thoroughly insulating one room, the management of the condominium sent a letter of termination, remarking that noise from sound systems was prohibited by the specific regulations of the building.

Lesson Learned: Always take into account all regulations; there are standards regarding so-called normal behavior in dwellings that set a rather low target for noise generation, be it music or anything else. More to the point, whenever any complaint appears, do react at once.

5.5.9 Grinding Stone

A small craftsman facility had been operating in the center of a large European city for centuries. The facility was located at the ground level of a 16th-century building that also included dwellings. Among the tools of the trade was a large grinding stone used for sharpening knives.

One day a new tenant took the upper floor (sixth floor) dwelling and quickly complained of the noise in her room that was generated by the operation of the grinding stone. As the emergence (exceedance over the background noise level) value was slightly above the legal target value, the craftsman was ordered to apply noise and vibration control on his equipment (which eventually required the mounting of the grinding stone assembly on an inertia mass; fortunately, the craftsman was the owner of his shop and basement, as significant structural work had to be undertaken to accept the extra weight).

Lesson Learned: Being there first (i.e., before the complainant) may not be enough of an excuse.

5.5.10 Spin Dryer

A cleaner had been installed for some time in a suburban house. When a new house was built and commissioned, the owner complained of heavy vibrations in his house, to such an extent that a glass could be seen slipping on the table during certain phases of the work.

Investigations showed that such vibrations were produced when the spin dryer was coming to a stop. Looking at the possible transmission paths, no structural connections were found between the two houses that were 10 m distant.

As the investigations had also hinted that the levels of vibrations inside the cleaner's house were a bit too high for its structural safety, it was quickly decided to have the equipment mounted on an inertia mass resting on resilient pads. This eventually solved the problem.

5.5.11 Artificial Valvula

One cannot make it more personal, we think! A young lady phoned the acoustic laboratory of the university and required guidance about the possibilities of noise control of her artificial valvula. On hanging up the phone, it was quickly decided that the simplest would be to show her through a sound level measurement in the lab that she was just internally hearing that noise and nobody else could hear it.

Well, when she came visiting the lab, we actually did discover that a clicking noise akin to a mechanical clock could be heard all over the place. Past the surprise, it was eventually noted that the laboratory featured a background noise level of 22 dB(A); with an artificially generated background noise level of 27 dB(A) (which is already much lower than in most public spaces), it was much harder to identify and she was then quite reassured. More to the point, we hastily cut out a noise shield worn as a plastron and made of heavy rubber-like material that helped reduce the body radiation, and she left the laboratory in a much lighter frame of mind.

Lessons Learned: Do not rely on your imagination alone. Wait for actual measurements and observations before deciding whether there is a problem. Next, use simple noise control solutions to solve it, and explain their reason to the user.

5.5.12 Laboratory Cabin

A laboratory working on human biophysics decided to examine over a week period such parameters as gases rejected or inhaled, as well as temperature fluctuations during certain tasks and their subsequent resting periods. It was deemed more expedient to have an industrial airproofed cabin custom modified for those purposes. However, on delivery it was found that the noise level inside the cabin was 80 dB(A)! This was of course totally unsuitable for most of the planned experiments. The contractor merely pointed out that he had delivered a cabin that was up to the industry's standards; furthermore, there was no acoustic clause in his contract. His argument was ultimately accepted by the court.

Lesson Learned: Always state the intended use in the acoustic specifications beside the acoustic objectives. Should you feel unequal to the task, then call a knowledgeable acoustic engineer.

5.5.13 Home Cinema

When they acquired their new flat, a middle-aged couple decided to allocate one room to their home cinema. A specialized contractor came and visited, and made a submission for the installation of the system, complete with the image and sound system and the required absorptive materials. When the room was duly fitted out, the owners were very happy with the image and sound. But not so were the neighbors, who complained that they had the sound without the image or any choice in the program! When questioned, the contractor merely shrugged and pointed out that the contract included the interior acoustic treatment of the room as well as the installation of the image and sound system, but it definitely did not include anything regarding extra acoustic insulation to other spaces. The owners were left with no other choice than bitterly moving out of the flat.

Lesson Learned: When acquiring a sound system, always make sure that the required sound insulation either is already available or will be delivered.

REFERENCES

1. European Framework Directive 89/391: *OSH Framework Directive on the introduction of measures to encourage improvements in the safety and health of workers at work*, June 12, 1985.
2. I. Fraser, *Guide for the application of European Directive 2006/42/EC machine directive*, European Commission on Industry, Luxemburg, 2010.
3. ISO Publication R *1996: Acoustics—Assessment of noise with respect to community response*, Geneva, 1971.
4. ISO 1996-3: *Acoustics—Description and measurement of environmental noise—Part 3: Application to noise limits*, Geneva, 2003.
5. FDIS ISO 1996: *Acoustics—Description, measurement and assessment of noise in the environment*, Geneva, 2014.
6. Recommandation du Ministère de la Santé du 21 juin 1963 relative au bruit (Recommendation of the French Ministry of Health on noise, in French), Paris, June 1963.
7. ISO 15667: *Acoustics—Guidelines for noise control by enclosures and cabins*, Geneva, 2000.

Chapter 6

Room Acoustics

6.1 INTRODUCTION

Room acoustics has been defined by a physicist as a mere subchapter of enclosed spaces acoustics that itself is a subchapter of continuous medium mechanics [1]. However, for many acoustically minded people, this is akin to the ultimate domain of acoustical engineering when thinking of performance halls. But production halls are also volumes to be contended with! And so are open-space offices.

Room acoustics mainly revolve around the same concern: enabling the occupants to hear the sound signal coming from the location of interest while not being unduly exposed to unwanted noise from other locations, ensuring proper sound propagation in the room, controlling the reverberation inside the room, and specifying the relevant noise criteria for the mechanical equipment.

This chapter is devoted to the basic physics of those phenomena. Application of those bases will be done in the relevant following chapters.

6.2 BASIC PARAMETERS

6.2.1 Foreword

Very early on the need to try to characterize the acoustics of a room had been understood. The composer Händel was known to clap his hands in the place where he was going to have a concert prior to deciding on the program; similarly, the composer and organist Silbermann would hit the floor with his cane to assess his acoustic environment. A significant step was spanned when a professor at a school of architecture named Sabine eventually noticed how the acoustics of his lecture hall would change according to the number of cushions brought in by the students! This led the way to the notion of quantity of absorption.

6.2.2 Acoustic Absorption

An acoustically absorptive material typically is either a porous or a fibrous material whose absorption coefficient is greater than 0.5 (cf. Section 3.5), meaning that a sizable part of the incident acoustic energy is physically absorbed and no longer active in the room. Physically, the vibrations of the air inside the fibrous channels of the material or the cells are turned into heat [2]. Examples of such materials are mineral wool, open-cell foams, and heavy velvet curtains.

Absorptive materials can be hidden behind a better-looking surface as long as it is not airproof, e.g., perforated sheet, fabrics, etc.

The use of absorptive surfaces is a tool that must be weighted: While an absorptive surface in front of the performers will prevent the occurrence of an eventual echo, it will also dampen the sound. This has a couple of consequences on the design of halls: one should not have a distance to the rear wall that may generate an echo (i.e., more than 17 m from stage to rear wall), and one should avoid too steep an audience in front of the stage.

An absorptive surface will prevent an acoustic reflection on it. This can easily be used in a theatre where one wants to avoid lateral reflections for the sake of intelligibility, but this usually must also be avoided when designing a concert hall where such reflections are needed for the sake of spaciousness.

6.2.3 Reverberation Time

6.2.3.1 Reverberation

Reverberation is the product of the multiple acoustic reflections on the inner surfaces of a room. In a well-balanced room those reflections will be well spread over the observation time span, with the listener unable to point out a specific reflection. But if several reflections arrive at the same time at the listener's ears without much acoustic energy around them, then one has an echo. To be discernable, this event must be at least 100 ms from any other event (this has a practical consequence: A reflective rear wall will not be positioned any further than 17 m from the stage).

The reverberation time, given the symbol T or RT, is defined as the time needed for the sound pressure level to drop by 60 dB from the moment the sound source is stopped. Its measurement calls for a sound source that is abruptly stopped. The choice of measurement points is described in standards, for example, ISO 3382-1 for ordinary rooms [3] and ISO 3382-2 for performance halls [4]. A simple evaluation can be carried out using Sabine's formula (cf. Section 2.6.2.4).

In order to better cover the aspects linked to articulation (for both speech and music), early decay time (EDT) has been introduced. EDT is defined as the time needed for the sound pressure level to drop by 10 dB from the moment the sound source is stopped.

The reverberation time was historically one of the first parameters to find its way in room acoustics. It is important when dealing with the comfort of both the listener and the speaker. However, it is not sufficient to characterize the acoustics of a room. For example, a RT of 0.9 s can be found in a small conference room that will be evaluated as suitable by its users, but it can also be measured in a fully fitted open-space office with a reflective ceiling where its users will feel the acoustic conditions are simply awful!

6.2.3.2 Echoes

The term *echo* comes from the name of a Greek nymph who was cursed to repeat what others had just said.

As pointed out earlier, in a well-balanced room acoustic reflections will be well spread over the observation time span, with the listener unable to point out a specific reflection. But if several reflections arrive at the same time at the listener's ears without much acoustic energy around them (that is, more than 100 ms away), then one has an echo. A simple typical occurrence will be in a hall with a reflective surface on the back wall. The listener will first hear the direct sound coming from the speaker, and then he will later experience the reflected sound coming from the back wall. Of course, the longer the distance, the higher the chance to experience an echo.

A more complex occurrence will typically be experienced in some end-of-the-19th-century performance halls, where a cupola over the audience was fashionable; more to the point, the ceiling over the proscenium often featured a concave shape. In addition, the back wall of such halls often was of the curved type. The net result usually is a strong echo, and sometimes a loss of spatial awareness (cf. Sections 6.10.4 and 6.10.5).

Flutter echoes result from the presence of two parallel reflective surfaces. When an impulsive sound is generated between those surfaces, one can hear a succession of impulses. In addition, one's voice turns metallic. This is not uncommon in some mistreated halls where acoustic absorption is found on the back wall (accounting for longitudinal propagation) and on the ceiling (accounting for vertical propagation), while lateral walls are kept bare (e.g., because of large windows) and parallel.

Focusing is the result of several reflections arriving at the same time at the listener's ears. When the direct sound is either no more than 22 ms away or lost in the background noise, such focusing is useful for natural acoustic transmission purposes. There are quite a few famous examples of such focusing, for example, St. Peter's Basilica in Rome, the whisper wall in Beijing, or the old vaulted subway stations in Paris.

6.2.3.3 Impulse Response

In simple nonmathematical terms, the impulse response is the sound level versus time curve when the room has been excited by an impulse sound (e.g., a clap). Nowadays, it can be processed from measurements and computer simulations using dedicated software.

The impulse response can bring useful indications regarding the behavior of a room that goes far beyond the mere reverberation time value. It will indicate the various contributions from reflections, and using the time interval between initial signal emission and reflection arrival, it will then be possible to determine the path followed by that particular contribution. While the impulse response as a measurement result will sometimes need some head scratching to find out which path has been followed, as a computer simulation result it usually is possible to follow precisely such a path and take appropriate measures if need be.

Figure 6.1 shows an example from a badly treated Italian-style opera hall. Looking at the stylized impulse response, it can be seen that after the direct sound has reached the measurement point, there are a few reflections, and then there is a massive arrival of reflected sound; this comes from the fact that such a facility features a curved cupola, curved balconies, and

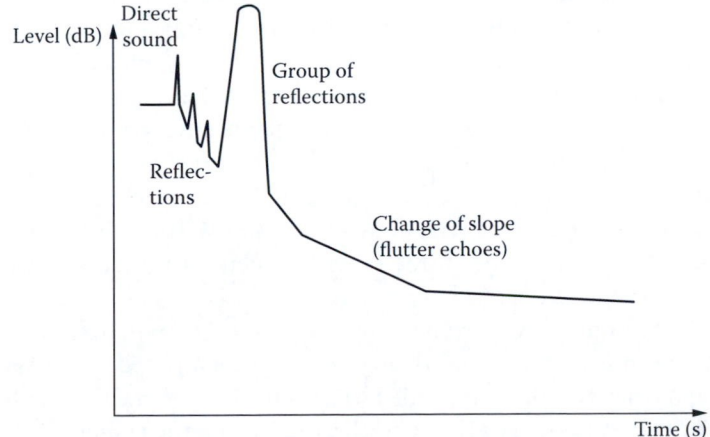

Figure 6.1 Example of stylized impulse response curve.

a curved proscenium, whose reflections arrive near simultaneously more than 40 ms after the direct sound.

6.2.3.4 A Few Practicalities

Clearly enough, there are a few basic rules to be followed in order to avoid a major problem.

To start with, concave surfaces are liable to create unwanted focusing and will probably create echoes. Their use should be closely monitored. Do remember that an absorptive treatment will not necessarily be the cure to any problem, as while it will eliminate the reflections under scrutiny, it will also reduce the reverberation time, and it will affect sound propagation too. More to the point, most fibrous absorptive materials will have an absorption coefficient value of at least 0.8 in the 2000 Hz band, while it will merely be only 0.1 in the 125 Hz band. This has a practical consequence: Attempting to reduce the reverberation time in the low-frequency range using fibrous materials only will reduce its value in the high-frequency range even more; thus, in order to achieve a reasonable reverberation time balance over the frequency range of interest, other means (e.g., a thin plate on an absorptive void) will probably be needed.

Next remember Sabine's theory: It is applicable in diffuse fields, typically in volumes whose dimensions are quite similar to one another. So this is another scheme to be avoided.

Should one combine similar dimensions and such a concave surface as a cupola, well one is history then! A famous example is that of a conference hall in the Fogg Art Museum of Boston, where such a scheme was applied to such an effect that the curators unanimously decided to have a plate at the entrance explaining why the name of the architect was given to this hall [5]!

6.2.4 Spatial Sound Level Decay

The reverberation time is not sufficient to characterize acoustic properties of a room to try to establish a comparison between rooms. In the above example of an open space, while the RT is perceived as correct, it is clear that the attenuation between workstations is not! In order to address this point, one can use the spatial sound level decay per doubling of distance, given the symbol DL_2, that gives a statistical indication of the diminution of the sound level value measured at a distance d from the sound source, compared to the one measured at a distance $d/2$. It is a good indicator of the precautions taken in the acoustic treatment of the room by means of absorptive surfaces (especially the ceiling) and eventual noise barriers. Procedures of measurement in an empty room are given in standard ISO 14257 [6], and in an open space in standard ISO 3382-3 [7]. According to the requirements of the investigation, it can be expressed in dB(A) per doubling of distance for a pink noise or for a speech spectrum. One should be careful when performing such measurements, as quite a few points may influence the result: To start with, the engineer's sound source seldom is totally omnidirectional, and at close quarters this will show (cf. Section 6.10.3). Next, the location of the measurement point with regards to obstacles will influence the result (either through extra acoustic reflections or due to a noise barrier effect).

In an enclosed space, going away from the sound source, one will first experience the direct sound field from the sound source, and then gradually be subjected to the reverberant sound field created by acoustic reflections around. The distance where the direct sound field is equal to the reverberant field is known as the critical radius, given the symbol r_c. One can write:

$$4\,(1-\alpha)/A = Q/(4\,\pi\,r^2) \text{ with } r = r_c$$

where α is the mean acoustic absorption coefficient in the room, A is the total absorption area of the room, in m², Q is the directivity factor of the sound source, and r is the distance from the sound source to the receiver, in m.

In the 18th century interest was shown in finding some quantification of the human voice with regards to its force and directivity. This eventually led to a front-directed ellipse equal contour, with the immediate consequence that a fan-shaped hall will not be as efficiently covered as a rectangular one. With a better understanding of acoustics, it was found that the direct sound field (i.e., the sound propagating directly from the sound source to the listener) can be perceived until a distance equal to three times the critical radius, r_c. This gives a fair idea of the maximum admissible distance, D_c, between the stage and the last row of seats:

$$D_c = 0.2\ \mathrm{sqr}(VQ/T)$$

where V is the volume of the room, in m³, Q is the directivity of the sound source (2.5 for a human mouth), and T is the reverberation time, in s.

Note: D_c is sometimes known as the critical distance, not to be confounded with the critical radius, as D_c is roughly 3.4 times the critical radius [8].

6.3 A FEW TYPES OF ROOMS

6.3.1 Performance Hall

Performance halls are spaces where the audience can listen to performers. In such spaces the RT has to be adjusted to the required value to enhance the performance (i.e., providing good intelligibility of the voice signal or strength). In addition, it may be needed to achieve spaciousness. The case of performance halls is examined in Chapters 12 to 15. According to the type of activity (and the history of the building too for an ancient venue), various shapes can be found; a few of them are sketched in Figure 6.2.

6.3.2 Industrial Hall

Industrial halls and workshops are spaces where noise from activities must be controlled. In such spaces the RT is often rather low due to the numerous fittings inside, but the spatial sound level decay must be adjusted to provide attenuation between workstations and prevent undue noise exposure. The case of industrial halls and workshops is examined in Chapter 10.

6.3.3 Open-Plan Office

Open-plan offices are nonpartitioned office spaces where one must be able to work without being unduly disturbed by the noise of other workers, especially speech signals. In such spaces the spatial sound level decay has to be adjusted to provide reasonable attenuation between workstations. The case of open-plan offices is examined in Chapter 7.

6.3.4 Meeting Rooms

Meeting rooms are spaces where a group of people convene a meeting in which they must be able to debate orally. In such spaces emphasis is put on speech intelligibility, and accordingly,

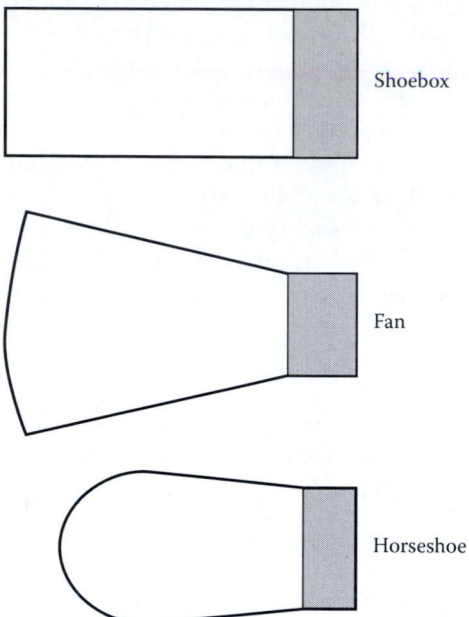

Figure 6.2 Examples of a few floor shapes of halls.

the RT is kept rather low, yet propagation must be ensured between the participants, which means it should not be too low either. The case of meeting rooms is examined in Chapter 11.

6.3.5 Classrooms

Classrooms are spaces where the students can listen to their teacher. In such spaces the RT is kept rather low to provide good intelligibility of the voice signal, yet it must not be too low; otherwise, there will not be correct propagation between the teacher position and the last row (typical regulations require a RT value in the 0.4 to 0.8 s range). The case of classrooms is examined in Chapter 11.

6.4 WHAT PARAMETER FOR WHICH USE?

6.4.1 Foreword

As pointed out earlier, the RT is often not sufficient to describe the required acoustics of a space. Some more parameters are needed, keeping in mind that quite a few parameters have been tentatively developed, but few have so far survived as well as the RT [9]!

Just as a reminder: Those parameters are computed or measured at a given point in the room; achieving a good value at one point may not mean that one will achieve it at each point in the room!

6.4.2 Spaciousness and Envelopment

Spaciousness characterizes the impression of a large and enveloping space.

Envelopment characterizes the acoustic impression of surround to the listener (as opposed to the impression of frontal sounds).

6.4.3 Clarity and Fullness

Clarity C_{80} characterizes the intensity of the direct sound, compared to the reverberant or reflected sound. It is defined as the difference, in dB, between the sound energy received by a listener in the first 80 ms and the remaining sound energy. The greater the reflected intensity, the more full the room is perceived and the less clear the spoken word or fast or highly articulated sections of music are perceived. Clarity is sacrificed in a "muddy" room, but a lesser degree of clarity enhances slow passages of music from the romantic era, where one prefers a global sound impression over separate pulpit contributions. To achieve greater clarity or definition, the entire audience should be close to the stage and have an unobstructed view. This can be accomplished by placing performers on a raised stage, and perhaps a raked stage, and by placing the audience on a sloped floor or in balconies.

Assuming an exponential sound decay in the room, clarity, in dB, can be estimated from the reverberation time using the formula [29]

$$C = 10 \log \{\exp(1.104/RT) - 1\}$$

6.4.4 Loudness

Loudness characterizes the average power detected by a listener. A longer reverb time will contribute to the loudness of the hall.

6.4.5 Strength

Strength, given the symbol G, at a given point of a room is defined as the difference between the sound power level L_w of the source and the sound pressure level L_p:

$$G = L_w - L_p$$

Using the room parameters, this can be expressed as

$$G = -10 \log \{(4\,(1 - \alpha)/A) + (Q/(4\,\pi\,r^2))\}$$

6.5 SPEECH INTELLIGIBILITY

6.5.1 Foreword

How do we characterize speech intelligibility? Initially, the usual method called for a panel of listeners who were supposed to listen to a list of meaningless words and write them down. Another test procedure called for full sentences to be read and written down by the panel. Of course, such tests are sensitive to the hearing ability of the listeners, but also to their cultural background. More to the point, it is not always easy and cheap to assemble a large enough group of listeners to perform such tests.

The next step was to try to devise a procedure based on measurements. It was quickly pointed out that speech intelligibility was dependent on two factors:

- Signal-to-noise ratio: The smaller this ratio, the harder the understanding (or even the awareness!) of a speech signal.
- Reverberation time: The shorter the reverberation time, the better the articulation (i.e., consonants) perceived by the listener.

Simple indicators could thus be accessed through a measurement or computation of the reverberation time and background noise.

6.5.2 STI and Derivatives

STI stands for speech transmission index. It was introduced in the early 1970s [10]. It is a numeric representation of communication channel efficiency, with a value ranging from 0 (bad) to 1 (excellent). In practice, one usually aims for a STI greater than 0.5 in most situations. It predicts the likelihood of syllables and sentences being comprehended, and is language independent. A somewhat simplified version, named STIPA, is available for PA systems.

CIS stands for common intelligibility scale [11]. It is linked to STI through the relation

$$CIS = 1 + \log(STI)$$

A CIS of 0.7 minimum is required by the European standard on PA systems for emergencies [12], which is quoted in the relevant European regulations. The same value is also requested by the National Fire Protection Association (NFPA) in the United States.

RASTI stands for rapid speech transmission index. This was a practical time-saving solution popularized by the measurement manufacturer Brüel & Kjaer when full STI measurements were practically complicated. Instead of operating in several frequency bands, it only considered the 500 and 2000 Hz octave bands. As of 2011, the International Electrotechnical Commission (IEC) has declared it obsolete.

6.5.3 SII

SII stands for speech intelligibility index. It is a numeric representation of communication channel efficiency, with a value ranging from 0 (completely unintelligible) to 1 (perfectly intelligible). It uses 21 critical bands (or 18 third octave bands) and enables experienced operators to diagnose causes of loss of intelligibility [13].

6.5.4 ALc

ALc, also known as Alcons, stands for articulation loss of consonants. As consonants are known to have more importance than vowels in speech intelligibility, it is fair to consider the percentage of consonants not understood as a measure of bad intelligibility [14]. It is quite simple due to its use of the 2000 Hz third octave band and was quite popular in the days where measurements and calculations could be lengthy.

The lower the value of ALc, the better the quality of speech transmission is. Its numeric value typically ranges between 0 and 10% for good intelligibility, while 15% is the maximum acceptable loss. A quick estimate of ALc, expressed in percent, at a distance much greater than the critical radius r_c can be obtained using the formula $ALc = 9\ RT$.

6.5.5 Electroacoustics

It is not uncommon to hear an architect or an end user wondering aloud why his beautiful architecture should be wasted for the sake of acoustics while it is so easy and unobtrusive to implement an electroacoustic system.

Well, sure enough an electroacoustic system can add acoustic energy, and it can of course be useful in certain situations. But it will not remove unwanted acoustic energy such as the one originating from an echo.

There are quite a few books dealing with electroacoustics (e.g., [14]), so this section will just remind the reader of a few practicalities. To start with, the electroacoustic system and the room must be suited to one another. If one cares to remember the basics of speech intelligibility, two main things must be taken into account:

- Make sure the reverberation time is low enough (in order to help recognition of consonants): This is pure room acoustics, as much as preventing the occurrence of echoes.
- Make sure the (speech) signal-to-(background) noise ratio is good (a target of 10 dB is usually considered acceptable for safety purposes): This is the task of the electroacoustic system, bearing in mind that the signal will be delivered by the nearest loudspeaker, but all the other loudspeakers will most probably act as noise sources.

A few schemes can be envisioned: The simplest one will call for a single regular loudspeaker at one end of the space and absorptive walls and ceiling. While this is pretty simple (one does not have to worry about the effect of other loudspeakers), it also means that in order to achieve a reasonable sound level value at the other end of the space, one may end up blasting the eardrums of people closest to the loudspeaker. In order to avoid such a problem, one may then be tempted to spread a few ceiling-mounted loudspeakers over the space. This may definitely work provided that the signal emitted by other loudspeakers does not come to the ears of the listener with too long a delay. In order to make sure of that, a delay device will probably be needed on the lines to each loudspeaker. Last, it may also be possible to opt for a "line array" set of loudspeakers. They behave like an acoustic antenna and, when properly studied, can do wonders.

There are a couple of things that must really be avoided (please kindly note this is not an exhaustive list—and yes such things have been seen!):

- Hiding the loudspeaker behind a solid or airproofed foil or plate (one will lose directivity and deform the spectrum)
- Pushing the emitted sound level to the limit (not only may the signal be clipped by the electroacoustic system, but also most of the loudspeakers will eventually turn up as noise sources)
- Spreading lots of loudspeakers without bothering with a delaying device (most of the loudspeakers will eventually turn up as noise sources)
- Directing the loudspeakers toward reflective surfaces (this will quickly increase the reverberant sound field and therefore the noise levels)
- Leaving open the holes used for the cabling of the loudspeakers (this will decrease the sound reduction of the relevant wall or ceiling—incidentally, it may also turn out to be a serious weakness for fire safety purposes too)
- Laying the electroacoustic system cabling close to power wires

Some electroacoustic systems have been devised to artificially enhance reverberation. Among such systems some can be designed to also be of use for regular sound diffusion improvement (e.g., [15]). Needless to say, such systems do need a specific study.

To sum up, electroacoustic systems are not a substitute for natural acoustics. A defect in the latter will find its way in the overall acoustic quality of the venue no matter what. But with a suitable study, they can enhance the acoustics of the venue, and of course they are invaluable for PA and safety announcements.

6.5.6 Articulation Index and Privacy Index

The articulation index (AI) gives an indication of the intelligibility of speech heard in a given noisy environment, with values varying from 0 (completely unintelligible) to 100 (completely intelligible). A graphic method of determination is given in reference [16].

The privacy index (PI) is based on the articulation index (AI) trough the relation

$$PI = 100 \ (1 - AI)$$

PI describes the ability to prevent the understanding of speech, with values varying from 0 (no privacy) to 100 (completely confidential).

Speech privacy class is defined by standard ASTM E2638-08 [17]. It applies to enclosed spaces.

Secret	PI = 100	Grade A+
Confidential	95 < PI < 100	Grade A
Normal	80 < PI < 95	Grade B
Transitional	60 < PI < 80	Grade C
None	PI < 60	Grade F

6.6 DIFFUSION AND SCATTERING

Specular reflections on a surface may be a cause of trouble, as they might eventually result in echoes. For example, two parallel reflective surfaces may easily result in a flutter echo; also, concave surfaces may generate a strong echo. In order to prevent such occurrences, it is of interest to introduce diffuse reflections instead of specular reflections. This can be performed through the use of asperities on the surface (Figure 6.3) in order to avoid the

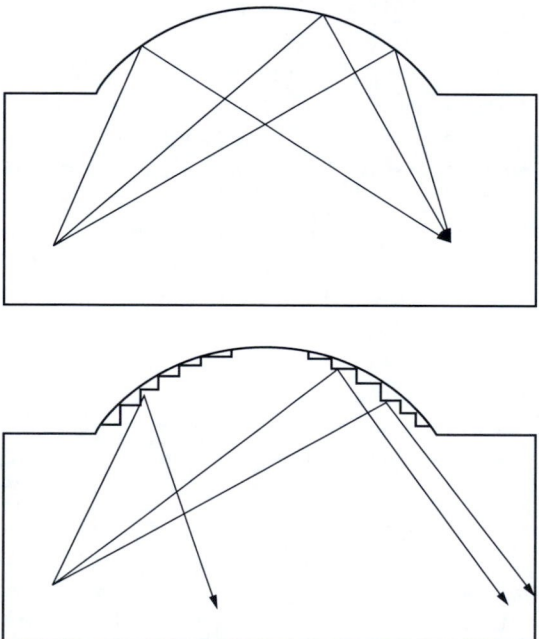

Figure 6.3 Effect of diffusive treatment on a concavity (e.g., a cupola) on the propagation of sound in the space beneath.

Figure 6.4 Example of diffusive finishes on a wall.

further use of acoustic absorption that would ultimately reduce the reverberation time value (cf. Section 6.10.10). Figure 6.4 displays a few examples of such surface finishes. The efficiency of the diffusion will depend on the size and shape of the asperities, and it is frequency dependent. Coefficients are used to characterize this acoustic performance. Presently there is a scattering coefficient in the diffuse field whose measurement procedure is given in standard ISO 17497-1 [18]. This coefficient characterizes scattering. It is in the range 0 (reflected energy is only specular) to 1 (reflected energy is totally diffused).

There also is a diffusion coefficient in the free field, whose measurement procedure is given in standard AES SC-04-02 [19]. This coefficient characterizes the spatial uniformity of reflections. It is in the range 0 (energy is redirected) to 1 (uniform spatial distribution).

On a more modernistic tone, Figure 6.5 displays an example of a quadratic residue diffuser, made of periodic wells and asperities [20]. The lower-frequency limit of such a diffuser is given by the well depth (that should be 12 times the wavelength of the lowest frequency), while the upper frequency limit is given by the well width (that should be half the wavelength of the upper frequency) [31].

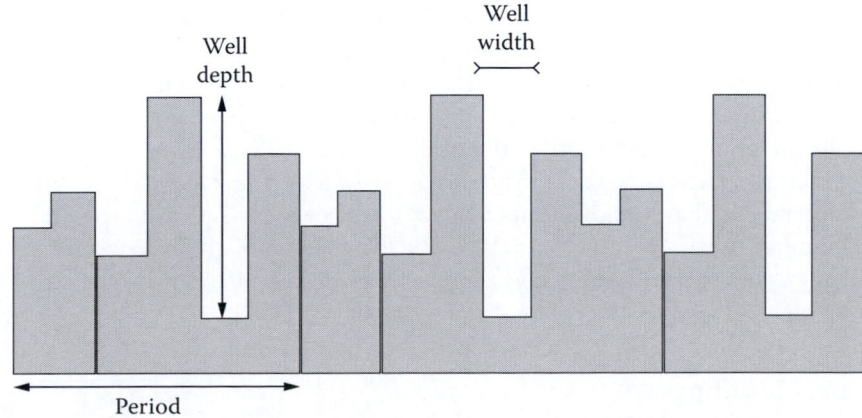

Figure 6.5 Example of quadratic residue diffuser.

6.7 MODELING

6.7.1 Foreword

The time when all rooms and halls were purely empirically built is long gone. Nowadays, a minimum of computation tools are available; all that is left to do is to properly use them.

To start with, one must be conscious that as implied with the title, we are talking of a model, that is, a representation of the actual physical phenomena. Modeling in acoustics is akin to freezing in cooking: If you start with good ingredients and properly perform the freezing, then later on the unfreezing, you will end up with at least an eatable, if not palatable, dish, but if you use rotten ingredients or mess up the process, the result will be awful. In acoustic modeling, if you start with bad data or forget about the validity domain of the model, the result will be messy, to say the least.

6.7.2 Statistical Modeling

Statistical modeling has been used for a long time. In the crudest models one simply needs the volume of the room and the areas of the surfaces together with their absorption coefficients.

Sabine modeling is of course the simplest. The sound pressure level L_p at a distance d from the sound source of sound power level L_w is given by

$$L_p = L_w + 10 \log [4 (1 - \alpha)/A + Q/4 \pi d^2]$$

where Q is the directivity of the source, α is the mean absorption coefficient, and A is the total absorption, in m².

6.7.3 Ray Tracing Modeling

Ray tracing using a computer does look the part. However, the physically minded can quickly understand that solving the wave equation inside a hall full of singularities, such as balconies, coupled volumes, and funny-looking surfaces (including curved surfaces), is not going to be easy, not to say accurate. Nevertheless, ray tracing is rather quick to implement (especially if the architect has previously been talked into providing the relevant drawings suitably clean of unnecessary details) and can be used to pinpoint and correct such defects as echoes and inadequate reverberation time. Some commercial programs now even provide auralization modules that enable the user to hear how the hall under study will eventually sound.

Good news: Some computer modeling software can now take into account diffusion. Now for the bad news: The diffusion coefficient value usually is taken on the basis of experience at 0.25 minimum for the less diffusing surfaces in order to try to get a rather diffuse field. More to the point, while in the real world an incident ray will result in a multitude of diffused rays, most of the time the software will only send a single ray in a random direction [29]. This means that a high number of rays are needed (as a rule of thumb, one may consider a number of rays greater than 10 times the volume of the hall expressed in cubic meters). Last, curved surfaces are not yet properly taken into account by the software [30], and focusing effects may not always be properly dealt with [29].

6.7.4 Scale Modeling

Scale modeling may look old fashioned, but it always features a kind of magic to onlookers. Actually, one may discuss scattering objects for hours in front of a dubious client using some computer simulation results and images without much success, but having the client look at a scale model while being explained the problem is usually quite effective, not to say awesome.

Among early attempts at scale modeling, work was carried out in France in the beginning of the 1940s by Canac [21] in the study of antique theatres using water as the medium of propagation.

Scale modeling proves efficient when dealing with singularities such as balconies, coupled volumes, and funny-looking surfaces. What scale should be used? Clearly, the smaller the scale, the higher the frequency range will be, and the more complicated the measurements will turn out, as air absorption becomes quite a problem. Scales like 1/10 to 1/16 are rather nice and popular, as air absorption is not yet too much of a problem, but while the model is easy to work with, it also happens to be quite cumbersome. Experiments have been performed with cardboard 1/50 models [22].

Scale modeling and computer modeling actually are quite complementary: While the former will usually not manage to come up with a good estimate of the RT value, it will nevertheless nicely display any focusing effect. On a high-stakes project, both modeling methods will be used.

6.7.5 As a Short Summary

A statistical model like the Sabine model will enable the user to easily evaluate the rough dimensions and characteristics of a room or hall. This is especially true for small volumes in which their three dimensions are of the same order of magnitude.

A ray tracing model can provide a good estimate of the reverberation time; it may point at unwanted focusing, but will be in difficulty when coupled volumes (e.g., narrow balconies) and complicated surfaces are featured.

Scale modeling enables one to take into account coupled volumes and diffusion, but it can be quite costly and cumbersome.

6.8 A FEW PRACTICALITIES

Berlioz was a famous French composer but also a critic and writer. He once wrote of the Paris Opera Garnier [23] that on asking the hostess for a seat, she answered: "Do you want a seat to hear or to see?" There are quite a few compromises that have to be solved between the acoustic goals, the stage requirements (including lighting), the structural requirements, and the HVAC requirements. Most of them will be examined in Chapters 11 to 15 of this book. However, the following points are common to all types of rooms:

- Make sure the sound insulation of the room is compatible with the protection of the activities inside (i.e., find out what the highest noise levels in the spaces around are, what the background noise levels inside the room must be, and determine the required sound insulation).
- Make sure the sound insulation of the room is compatible with the neighborhood (i.e., find out what the highest noise levels in the room are, what the background noise levels in the neighborhood must be, and determine the required sound insulation).
- Leave enough space for mechanical services (the smaller the allocated space, the noisier the HVAC system).
- Determine early enough what the minimum dimension requirements are (e.g., for similar seating capacity, a concert room will feature a higher volume than a theatre). Remember that in addition to the audience area and the structural wall thickness,

there will be extra requirements regarding the thickness of claddings and the minimal width of aisles (the latter for both operating and safety purposes).

- Prepare for a potential fight between the acoustician and the architect and the stage engineer: The latter will look for a low enough lighting bridge that the architect will probably want to hide by lowering the ceiling, while the acoustician will usually prefer an unobstructed transfer between the stage volume and the audience volume, especially for music-oriented rooms.
- Prevent echoes: This will mean avoiding whenever possible concave shapes, and using diffusive or absorptive materials.
- Adjust the reverberation time according to the needs. In order to be able to find low-frequency absorption, some space (at least 20 cm) will be needed.
- Air lock-mounted doors are an efficient solution to the problem of noise transmission control through the doors as well as a way to avoid light intrusion during the show. In addition, there often are regulations limiting the acceptable effort to be applied to emergency doors that will limit the sound reduction value to be reached, which means that a single door is not efficient enough for acoustic purposes.
- Make sure as early as possible that the requirement for air lock-mounted doors has been understood. In case of emergency, it can be quite tempting for HVAC purposes to allow fresh air in the room through an emergency exit, that is, until the firemen point out that they will not allow air lock-mounted doors for such purposes (for fear of one set of doors left closed), and one will be left with a single door with disastrous acoustic results (remember too that safety requirements will usually put a rather low limit on the effort to be exercised to open an emergency door: this will drastically reduce the potential acoustic efficiency of the door, as the seals will not be properly compressed).

6.9 RENOVATION

Renovation is a quite usual procedure in Europe [25]. There may be quite a few reasons to go for it. To start with, older buildings usually feature a higher potential occupation ratio per ground square meter than the one allowed by normal recent constructions. More to the point, it helps the urban planners to keep a homogeneous urban appearance. Last but not least, it helps save time if the walls and floors (and sometimes even the roof) are kept; administratively speaking, it may also speed up things, for when the envelope of the building is kept, one often will solely require permission for fitting out the building spaces instead of a full building permit. In the case of performance halls, there often is a sentimental and historical value associated with the building that makes it hard to forget [26].

On the negative side, it is not uncommon to lack some suitable space to route the ducts and pipes through. Quite often, the walls and floors have inadequate acoustic performance and must be upgraded using plasterboard and mineral wool stud-mounted elements that will reduce the available floor dimensions and height. Last but not least, the available volume is not always suitable for the wished-for performances [27].

In such a kind of project it is necessary first to perform a diagnosis of the existing building in order to find out how it is built and where the sensible points are. Each specialty will have to perform its own diagnosis. One should remember that a structural diagnosis usually can be a bit destructive, as the structural engineer will typically cut through floors and walls in order to assess their composition, so the acoustician must make his measurements before this is done.

Note: There is a strong need for coordination between the interested parties, as depending on the planned sketch of the future rooms inside the building, some zones will be more sensible than others. More to the point, everybody must be aware of each other's needs.

In addition, one must be particularly aware of the safety requirements applicable: Whenever derogation is delivered, it will certainly entail so-called mitigating measures that will probably be costly in terms of investment or extra personnel (cf. Section 6.10.8).

The diagnosis through measurements will feature:

- Sound insulation measurements between rooms when the walls are kept. Note: This measurement will help assess the potential flanking transmission by those walls, as well as the potential sound insulation of those rooms.
- Sound insulation measurements and impact sound measurements between rooms at different floors. Note: This measurement will help assess the potential sound insulation between floors.
- If applicable, vibration measurements on the floors or walls. Note: This will help assess the eventual noise generated by vibrations (e.g., from rail lines nearby) and vibration levels inside the building.
- An acoustic diagnosis of the site will be performed as per a regular new construction project (i.e., assessing the sound level values on the site and finding out what the potential noise sources around are, as well as the potentially sensible zones around). Note: Do keep in mind that usually a renovation project will entail some demolition (now politically correctly described as unbuilding) prior to the actual construction work. Under the nice words one can already hear the concrete breakers hammering away, so one had better have a good look at the location of the nearest neighbors, especially those who are structurally linked to the building. It will probably be necessary to explain to the neighbors the basics of the project and point out that while some phases of the work will be noisy, they will be kept to a minimum of duration and their time schedule will be adapted while appropriate noise reduction measures will be implemented.

It must be stressed that the diagnosis will constitute the testimony to the acoustic performance of the building prior to any work. It is not only a basis for the acoustic studies (from which predictive computations will be elaborated), but it is also often a compulsory step to be able to ultimately prove that the initial acoustic performances of the building have not been compromised [25].

In the particular case of historical buildings, things can get quite complicated, as usually the façades and even the roofs must be preserved. In some cases it is even necessary to preserve some interior spaces (e.g., because of paintings on the walls or ceilings). Under such circumstances the acoustic objectives must be adjusted on a case-by-case basis, and specific solutions must be elaborated (e.g., introducing intermediate spaces around in order to prevent direct transmission to other spaces of interest, or working on the other side of the partition or floor using such doublings as a floating floor, a plasterboard ceiling, or half wall, with mineral wool in the void).

A special mention must be made regarding performance halls: Those are usually considered a historical landmark, and the end user may wish to preserve or even improve much more than the sound insulation characteristics (though it is not always the case, please see Section 6.10.7). This means that specific room acoustic measurements must be carried out in order to explain the physical phenomena that are behind the acoustic characteristics, and then questions must be addressed to the users to know for sure whether they actually want them to be kept as such. Only after that will it be possible to prescribe the relevant constructive solutions.

6.10 EXAMPLES

6.10.1 Intelligibility Test Gone Wrong

On commissioning a large church, an acoustician decided to have intelligibility tests performed; a listener's panel was duly created. To the acoustician's alarm, the results were quite poor, as only a third of the tests proved satisfactory. However, after interviewing each member of the panel, the acoustician eventually found that the same proportion of listeners were illiterate!

Lesson Learned: When using a human panel, always beware of the cultural background!

6.10.2 Classroom Mistreated

On building a small school, an architect called an acoustician to define the acoustic treatment of the classrooms. Accordingly, the acoustician typically specified a suspended ceiling with an acoustic absorption coefficient of 0.7. Shortly after commissioning, a baffled architect called back the acoustician: The classrooms were highly reverberant. It turned out that the architect had picked up a suspended ceiling on its good look, and unfortunately, its absorption coefficient value was only 0.1 in the octave band of interest.

Lesson Learned: A suspended ceiling is not necessarily an absorptive ceiling!

6.10.3 Measuring the Spatial Sound Level Decay

Let's first consider a measurement in an empty room. Measurements are performed using an omnidirectional sound source located at least 3 m from the nearest wall according to ISO 14257. The measurement points are located at a distance from the sound source that is a multiple of 2 and 3 m; they are taken at a height of 1.2 to 1.5 m from the floor and must be located at least 3 m from the nearest wall.

Let's first measure starting from the sound source: Write down the relevant sound level value for each point. When finished, turn the page and repeat those measurements going back to the sound source. If the same results are found at 2 and 3 m from the sound source, either you cheated or your luck is uncanny: Taking into account that the floor covering may not be reflective, that the sound source may not be as omnidirectional as advertised, and that the measurer is not that precise in his positioning, one should not be surprised to find discrepancies of up to 3 dB at those points.

Let's now consider a measurement in a fully fitted room (e.g., an open-plan office). Measurements are performed using an omnidirectional sound source located at a workstation according to ISO 3382-3. The measurement points are located at a distance from the sound source that is a multiple of 2 and 3 m; they are taken at a height of 1.2 m from the floor. You will find out quite quickly that according to the positioning of the sound source with regards to its environment (e.g., screen, tablet, etc.), the results may differ.

Lesson Learned: Always note all the measurement conditions.

6.10.4 An Art Deco Theatre and Opera

The Opera of Vichy, which had been dreamt by French emperor Napoleon III, was eventually commissioned in 1936 and seats 1500. It is sometimes referred to as a Vienna opera in miniature. It features a curved reflective back wall, but fortunately there are boxes to reduce focusing there. There also is a cupola over the audience and a concave ceiling over the proscenium. The operators knew enough to avoid using the seats of the first balcony due to the strong echo. When a full acoustic diagnosis of the hall was performed [24], this echo was readily identified, with a time difference of 40 ms over the direct sound. More to the point, the acoustic spatial perception of the stage was affected to such an extent that when a soloist singer would act on the right side of the stage, many a spectator would turn his head to the opposite direction, as there were so many reflections coming from the back wall.

6.10.5 A Concert Hall with an Echo

Due to its shape featuring concave back walls and a dome, the Royal Albert Hall in London was often jokingly pointed out as the only place where a mediocre music piece could be heard twice!

6.10.6 A Simple Flutter Echo Occurrence

Let's consider a covered gallery outside: There is a small concrete ceiling covering a concrete floor, there are no side walls, and the end walls usually are far away. When one hits the floor there is a regular succession of impulses to be heard, as the reflections from the sides are inexistent and the reflections from the end walls are too weak to be perceived. This is the simplest example of a flutter echo.

6.10.7 The Olympia in Paris

The Olympia in Paris is a well-known facility for musicals and pop concerts where many a French celebrity launched his or her career. When a major urban renovation scheme was implemented on the corresponding area, which also included a hotel, offices, and a theatre, it was found economically more interesting to dismantle the existing facility, demolish the existing building, and rebuild another one 30 m further, as this enabled more lucrative office surface to be built. However, due to the reputation of the Olympia, it was decided that the new facility should be identical to the old one.

An acoustic diagnosis was performed. It included numerous reverberation time and acoustic strength measurements in the facility, as well as some background noise measurements. It was found that the underground metro line generated noise in the facility too. A computer simulation model was made to understand the acoustic performances and help define the specifications for the new facility.

While the client eventually decided that the facility could afford a box-in-box construction to prevent noise generated by vibrations from the underground metro from being generated in the facility (and also to prevent noise from the activities in the facility from being transmitted to the expensive offices around too!), the few improvements that were proposed regarding the internal acoustics of the facility (including the zone under the balcony) were refused on the grounds that it would not sound as the original. Actually, the quest for the original appearance was pushed to the point where impacts in the wall made by some famous artists were recreated.

Lesson Learned: What the client wants, the client gets!

6.10.8 The Theatre of Douai, France

When the acoustician was called to visit the Theatre of Douai in northern France, which used to be closed for most of the time except briefly in summer, the director sarcastically welcomed him, pointing out that this facility was the last of a series of 30 similar facilities, as all the others had eventually burned down! It turned out that during the brief operating periods, no less than a dozen firemen were required to be on site. As the director was quite happy with his theatre, the diagnosis characterized its acoustics through reverberation time and acoustic strength measurements, as well as background noise measurements, in order to be able to preserve its basic acoustic characteristics. More to the point, sound insulation measurements were performed inside the facility.

The project called for the installation of modern fire detection and prevention techniques, as well as the installation of a suitable ventilation system. Sound insulation with regards to other spaces was improved, and the internal acoustics characteristics were kept through the use of suitable materials and seating.

6.10.9 Using Electroacoustics

A public venue featuring several public spaces had been fitted with an electroacoustic system. The basic idea was for a distinguished speaker to be able to make a speech in one of the main spaces, with his speech being broadcasted in this space, but also in other spaces, if so wished.

Right from the start the users were unhappy with the system: Whenever it was switched on, an electric humming sound was heard, distortion was occurring, and in some rooms there was a distinct feeling of unequal sound distribution. In a large room the users joked that speech intelligibility was worse with the system than with natural voice. In a meeting room with a seating capacity of 20, a system of microphone and loudspeakers had been fitted at each seat, but due to the reflective surroundings and absence of delaying device, the result was simply awful.

An acoustician was eventually called to look into these matters. He was in for a few surprises. To start with, the "absorptive" treatment of the spaces on which the sound contractor had based his submission was a low-grade spray-on treatment that did not manage an absorption coefficient greater than 0.3. Next, the sound system cabling had been installed along the power cables (which incidentally is not allowed by the technical guidelines), and that explained the humming electrical sound. In the largest space, the absence of a delay device had resulted in most of the loudspeakers being a mere noise source for the listener. Indeed, the measurements showed that speech intelligibility was much worse with the system than with natural voice! In a smaller space that nevertheless featured four loudspeakers, those were found mounted in pairs and unfortunately not in phase.

In order to try to solve that mess, the acoustician first had to prescribe some extra absorptive treatment in order to reduce as much as possible the reverberant field. Next, the whole system had to be recabled, this time using suitably proper cables laid away from the power cables and properly connected. The loudspeakers' phase was checked, and delaying devices were implemented where needed.

Lesson Learned (at cost): Do not attempt to install a sound system without a proper study.

6.10.10 A Curved Back Wall

The Maurice Ravel Auditorium in Lyon can pride itself to have been one of the first pre-constraint concrete structures erected in France. Unfortunately, it also claims some serious defects as a concert hall, among which the volume is quite small for such purposes, and initially the back wall was curved and reflective, which resulted in a tremendous echo.

When significant sound absorption was eventually added to the back wall to prevent the occurrence of a very strong echo, of course the reverberation time value in the hall dropped even lower. This was eventually compensated through the use of a reverberation enhancement system.

REFERENCES

1. M. Bruneau, T. Scelo, *Fundamentals of acoustics*, ISTE (Hermès), 2006 (1998).
2. J.F. Allard, N. Atalla, *Propagation of sound in porous media: Modelling sound absorbing materials*, J. Wiley & Sons, New York, 2009.
3. ISO 3382-2: *Acoustics—Measurement of room acoustic parameters—Part 2: Reverberation time in ordinary rooms*, Geneva, 2008.
4. ISO 3382-1: *Acoustics—Measurement of room acoustic parameters—Part 1: Performance spaces*, Geneva, 2009.
5. B. Katz, E. Wetherill, Fogg Art Museum Lecture Room, a calibrated recreation of the birthplace of room acoustics, presented at Forum Acusticum 2005 Proceedings, Aachen, 2005.
6. ISO 14257: *Acoustics—Measurement and parametric description of spatial sound distribution curves in workrooms for evaluation of their acoustical performance*, Geneva, 2001.
7. ISO3382-3: *Acoustics—Measurement of room acoustic parameters—Open plan offices*, Geneva, 2012.
8. Peutz Group, Internal courses, 2014.
9. J. Bradley, Proceedings of the Vancouver Symposium on Acoustics and Theatre Design, 1986.
10. T. Houtgast, H.J.M. Steeneken, Evaluation of speech transmission channels by using artificial signals, *Acustica*, 25, 355–367 (1971).
11. P.W. Barnett, R.D. Knight, The common intelligibility scale, in *Proceedings of IOA*, vol. 17, part 7 (1995).
12. EN 60849: *Sound systems for emergency purposes*, Brussels, August 1998.
13. ANSI S3.5-1997: *American national standard methods for calculation of the speech intelligibility index*, 1997.
14. M. Rossi, J. Jouhanneau, *Notions élémentaires d'acoustique, électroacoustique—Les microphones et les haut-parleurs (Basic notions of acoustics and electroacoustics—Microphones and loudspeakers*, in French), Lausanne, Tec et Doc, 2007.
15. M. Asselineau, Acoustique active et acoustique passive—Approches complémentaires ou opposées? (Active and passive acoustics—Complementary or opposed ways? in French), presented at CFA2010 Proceedings, Lyon, 2010.
16. H.G. Mueller, M.C. Killion, An easy method for calculating the articulation index, *Hearing Journal*, 43(9), (1990).
17. ASTM E2638-08: *Standard test method for objective measurement of the speech privacy provided by a closed room*, 2008.
18. ISO17497-1: *Acoustics—Sound-scattering properties of surfaces—Part 1: Measurement of the random-incidence scattering coefficient in a reverberation room*, Geneva, 2004.
19. AES SC-04-02: Paper AES-5id-1997 (revised 2009).
20. M.R. Schroeder, Diffuse sound reflection by maximum length sequences, *Journal of the Acoustical Society of America*, 57(1), 149–150 (1975).
21. F. Canac, *L'acoustique des théâtres antiques (Acoustics of antique theatres*, in French), CNRS, Paris, 1967.
22. M. Barron, A.H. Marshall, 1:50 scale acoustic models for objective testing of auditoria, *Journal of Sound and Vibration*, 77, 211–232 (1981).
23. H. Berlioz, *Mémoires*, CNRS, Paris, 1980.
24. M. Serra, Diagnosis of a historical performance hall—Case study, presented at Proceedings of Acoustics 08, Paris, 2008.

25. M. Asselineau, The challenge of heavy rehabilitation projects—Case studies, presented at ICSV13 Proceedings, Vienna, 2006.
26. Peutz, Report on the opera of Lyon, Paris, 1993.
27. Peutz, Report on the Toulouse Capitole orchestra pit, Paris, 2005.
28. Peutz, *Acoustics by Peutz—Theatres and concert halls*, The Hague, 2010.
29. A. Gade, *Acoustics in halls for speech and music*, Springer Verlag, New York, 2007.
30. M. Vercammen, The reflected sound field by curved surfaces, presented at Proceedings of Acoustics 08, Paris, 2008.
31. M. Long, *Architectural acoustics*, Academic Press, 2006.

Chapter 7

Offices, Open Spaces, and Restaurants

7.1 INTRODUCTION

Offices, open spaces, and restaurants acoustics mainly revolve around the same concern: enabling the user to perform a task while avoiding him or her being unduly bothered by the noise generated around. This may require the possibility of holding normal voice discussions within small groups while avoiding too much interference with similar groups or individuals. Of course, under such conditions some speech will be understandable, especially when the background noise level is not too high, and it is up to the acoustician—according to the client's wishes—to decide how much of it should be permitted to be understood. This actually sets the tone of the problem: intelligibility between tables, be they workstations or a restaurant.

One may have to consider spaces that provide privacy, or even security, but one must then keep in mind that such spaces will then require a high-performance envelope.

Apart from the acoustic requirements, such spaces are also often meant to project a corporate identity, which may reduce the potential actions by the acoustician. More to the point, such projects are often sensitive to budgetary aspects.

7.2 REQUIREMENTS

7.2.1 Applicable Standards

There are standards pertaining to offices. Some of those standards are concerned with the ergonomics of the workstation (e.g., how much lighting of the desk, what dimensions of the desk and chair should be chosen) and usually include a few lines regarding the acoustics of the workstation [1], though those lines were usually not drafted by acousticians. Other standards specifically cover the acoustics of offices [2].

7.2.2 A Few Points to Be Considered

7.2.2.1 Offices

As for any project, everything starts with a program. Basically, there is a cost objective and an available floor area in which a given number of people must be accommodated. In order for the project to be acoustically effective, one must first assess what the user's needs are. Typically, the following points are likely to be found [3, 4]:

- Several workstations have a need for privacy (e.g., director office, accounting, director meeting room); in addition, they will need an enclosed space for security reasons.

- There are individual workers who do not need frequent communication with their colleagues but who need to work without undue disturbance.
- There are some small teams whose members interact with each other: While there must be a good intelligibility within the team, other personnel (including those from other teams) should not be unduly disturbed by the frequent talking. Incidentally, the scenario in the restaurant is similar to the one of the open-space office, with the guests' tables equivalent to a team workstation.
- There are also some larger teams occupying a full floor. The problem is then far more complex: While such large teams are often subdivided into smaller units, they usually listen to whatever verbal or simply auditory signal appears; while one does not want undue disturbance from such signals, one does need to perceive them correctly. A good example is the dispatching center (e.g., emergency dispatching, control tower, etc.) where the incoming phone call will be treated by a dedicated worker at a specific workstation while other workers attend to their normal duties, but everyone will be listening intently in order to assess the relevant available assets to be activated. So, there simultaneously is a need for intelligibility and discretion!
- In addition, there often are some potentially noisy pieces of equipment (e.g., computers, printers); last but not least, unless proper precautions have been taken, there might be people using their mobile phones.

7.2.2.2 Restaurants

Standards pertaining to the acoustics of corporate restaurants are usually taken out from the standards applicable to offices' auxiliary spaces (e.g., [2]). Looking at a restaurant, one can definitely find a similarity with offices, from individual eaters to small groups and even large tables. Once again, the aim of the exercise is to enable people seated at a table to talk normally without undue hindrance from other groups. While there is no office equipment, there definitely are a few pieces of equipment around, such as the refrigerated displays, the ice or coffee machines, and the exhaust from the preparation area. On top of that, one may have announcements (e.g., "number X, your steak is now ready") as well as music. Last but not least, there may be some noise coming from the dirty tray conveyor and the cleaning station.

7.3 ACOUSTIC TARGETS

Unsurprisingly, the problem of internal acoustics will revolve around the question of intelligibility.

As usual, the project starts with a program: What does the end user (and the payer too!) actually want? This means the various requirements must be identified and the relevant acoustic objectives stated. In today's spirit of sustainable development, one must be conscious that a team effort (i.e., architects, structural engineers, HVAC, and acoustics, just to name a few) occurs and make sure that the various solutions that are considered at the design stage are compatible with each other. Acoustics often is part of a global problem that can only be solved by a complete design team fed the relevant data by the end user.

7.3.1 Sound Attenuation and Sound Insulation

Regarding the sound insulation D_{nTw}, there are a few values worth remembering:

- 35 dB is the minimum value to allow similar activities to be carried out in two offices without unduly annoying each other; it also is a value where it will be possible to hear the phone call next door if attentive to it.
- 40 dB is the minimum value to allow slightly different activities to be carried out in two offices without unduly annoying each other; it also is a value where it will be possible to hear a good part of the phone call next door if paying attention to it.
- 45 dB is the minimum value to allow privacy—as long as one is using a normal voice—between two offices. But it will nevertheless be possible to understand part of the conversation if a raised voice is used or if the listener is intently eavesdropping with his ear to the partition.
- 55 dB is the minimum value to allow security of speech (meaning even with the occasional raised voice or the listener's ear to the partition, it is not possible to understand the discussion).

Such values are, of course, sound insulation values, meaning one is talking of enclosed office rooms. With an open space there is no hope for privacy; the only thing one can aim for is discretion, meaning that the occasional discussion at a workstation will not unduly disturb the other workstations.

Just as a reminder, the attenuation between workstations in an open space is in the range 4 to 11 dB according to the distance and the eventual presence of barriers. The presence of reflective surfaces nearby will also be a deteriorating factor.

7.3.2 Noise Levels from Activities

When aiming for a speech transmission index (STI) of 0.2 at the nearest workstation located 5 m away, with a background noise of 35 dB(A), it turns out that a 20 dB attenuation is required. This is clearly out of the possibility of a noise barrier in an open space, so the background noise level value has to be increased.

As a guide rule, it may be considered a radius of distraction defined as the distance from which the STI will fall under the 0.50 mark [6].

As a reminder, the background noise level must be high enough to act as masking noise, but it must also be not so high that it would be tiring. The usual value is to be found in the range 38 to 45 dB(A) [5], though on the basis of experience, the 38 dB(A) value recommended for some sustainable development projects with spaces to be fitted out [7] is too low to enable efficient masking. Other recommendations sensibly impose both a minimum and a maximum noise level value [8].

One also has to take into account the noise from activity around: It is generally considered [9] that complaints will appear in open spaces with a L_{Aeq} of 55 dB(A) or greater. This is rather logical, as it is admitted [10] that phone conversations start being difficult with such a level, while normal voice conversations will be difficult with a L_{Aeq} of 65 dB(A) or greater. More to the point, WHO recommends not to exceed 55 dB(A) in offices [11].

The noise level pattern in a badly treated restaurant is striking: On opening the facility, the noise levels climb steadily until it is too tiring to speak or listen. The levels will then decrease until eventually people start realizing that it is again possible to talk, and the whole cycle starts again. On the basis of experience, the so-called cocktail party effect will start appearing between 65 and 70 dB(A) in such places.

Of course the presence of noisy pieces of equipment (e.g., printers plotters and photocopiers in an open-space office, or refrigerated shelves and exhaust fans in a restaurant) will exacerbate the problem, as one will be compelled to speak louder; this will in turn push other people around to speak louder too.

Attempting to enhance acoustic comfort in an office or a restaurant often means reducing the intelligibility of speech messages from other tables; this means it has to be assessed [12]. In order to achieve that feat, one may consider the use of masking noise. But what should be done when there is not enough of it? Artificial masking noise has steadily found its way into office environments as the lesser evil between annoying noise and noise [13]. A word of advice here: To start with, while it seems widely accepted in North America (perhaps because of the high noise levels already generated by HVAC?), to the point where there are standards covering the subject [14, 15], it often is warily viewed by potential users in Europe. There are two conditions for success: First, the average user should not be aware of the existence of such a system in the premises. Next, the noise spectrum used should be as close as possible to a noise criteria (NC) curve; more to the point, the sound sources must be numerous enough that the average user cannot pinpoint the location of such a source in the ceiling void. Last but not least, the sound level should not be too high (45 dB(A) really being a maximum).

Remember: If the average user discovers the existence of the masking system, everything will turn sour in no time. But using a system imitating ventilation noise may help [16].

7.4 A FEW BASIC RULES

Two stages are usually found in either office or restaurant construction. The first one is the erection (or perhaps rehabilitation) of the building shell, and it will be treated as a normal building operation, looking at the sound insulation with regards to the outside (either to protect the neighborhood from the noise from activities, or to prevent the intrusion of external noise, e.g., traffic noise, inside the premises) and investigating the mechanical equipment (both for comfort inside the premises and for community noise prevention purposes). This first phase of construction usually is carried out by a standard building contractor under instructions from a regular design team. However, there usually is a second stage where the end user, usually assisted by a space planner, will have the space fitted out to his needs. During this stage, some extra equipment will be added: An office may need an extra emergency generator and dry coolers to support its computer system, while a restaurant will need a large freezer as well as several exhaust systems. As that equipment will usually end up either on the roof or in the basement, in locations prepared for such purposes, there definitely is the matter of the definition of appropriate noise control measures, both for noise control inside the premises and for community noise control (with usually a tough question to be solved: Who, the builder or the fitter, is supposed to shoulder the associated costs?).

There are guidelines available in the relevant standards regarding minimal distances and recommended fittings for open spaces. It makes sense to read them prior to being committed to a given space planning project.

The Finnish Association of Civil Engineers (RIL) has published a standard on the acoustics of offices [25]. It recommends minimal values for the rate of spatial sound level decay DL_2 and the discretion radius r_D. A computer simulation software offering direct comparison between the predicted values and the recommended values is available on the Finnish Institute of Occupational Health (FIOH) website [25].

It is quite tempting for the operator or the end user to put as many people as possible within the available floor surface. This is, of course, true of quite a number of open spaces. But it is also true of many restaurants, either for economical reasons (e.g., the owner wanting his investment to pay back) or for service reasons (e.g., a given number of people must have finished their lunch by a given time; this can be the case of corporations as well as schools). Beware that under such conditions it will not be possible to exercise efficient noise control measures, as people will always be in the direct field of at least one talker.

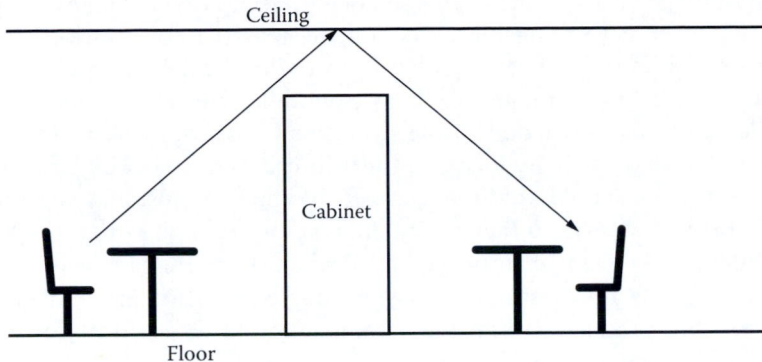

Figure 7.1 Effect of a reflective surface (e.g., ceiling) on the propagation of sound between workstations.

Remember, the largest potential reflector or absorber will be the ceiling. It does make sense to have a really absorptive ceiling over the whole area.

Should an area be earmarked for eventual partitioning using removable partitions, one must make sure that both the normalized sound level difference of the ceiling (and that of the technical floor, if applicable) and the sound reduction of the removable partition are capable of satisfying the acoustic targets.

Should partitioning be carried out using removable partitions, there is no hope to achieve acoustic privacy between the partitioned spaces.

Do not forget that lengthy or noisy phone calls, or impromptu meetings, can occur. A few enclosed cubicles (and the instructions on how to use them too!) are needed for that purpose.

The use of absorptive surfaces as close as possible to the scene of action is needed in order to prevent noise propagation throughout the office space. This means that the ceiling must be absorptive (i.e., the average absorption coefficient of the ceiling must be at least 0.60, including the lighting and ventilation terminals), as illustrated in Figure 7.1. But this is not sufficient: The presence of vertical reflective surfaces such as partitions and closets will significantly increase propagation. This is especially true of the large areas made by the façade or by the glazed partitions of enclosed spaces. Closets and nontransparent partitions can easily be clad by an absorptive material, and some manufacturers have such cladding in their catalog (e.g., [17]). Unfortunately, glazed areas cannot be turned from reflective to absorptive, unless using a microperforated sheet in front of them (e.g., [18]). This means that absorptive noise barriers will probably have to be used in the vicinity.

The use of barriers must be considered with care. They may constitute a hindrance for natural lighting, and even for a quick getaway. If high (which is better for acoustical purposes), they may interfere with ventilation and fire detection. Last, if they are reflective, they will merely redirect acoustic energy elsewhere. Guidelines about noise barriers are given in standard ISO 17624 [19]).

In addition, they may prevent the sighting of the whole space from the entrance, and this last point makes them especially unpopular with the restaurant operators and clients alike, as it becomes difficult to spot empty seats.

Experience shows that an absorptive barrier with a height (from the floor) in the range from 1.2 to 1.5 m usually is a good compromise between the various requirements.

Barriers can be used to achieve separation by entities or groups. When observing an office space, the existence of territories usually quickly appears [20, 21], and barriers may help regulate them, in addition to adding extra absorption in the space.

Background noise plays a significant role. As one is essentially dealing with speech intelligibility, in order to achieve a sizable reduction of signal-to-noise ratio, it is possible to raise the background noise level as long as it stays under the 45 dB(A) mark [5]. Actually, some standards definitely recommend the background noise level value not to be under the 40 dB(A) mark [22]. This can be done using either the HVAC or masking sound generators. Regarding the latter, they often are located inside the ceiling void, facing up to the underside of the concrete slab; one should kindly remember that they are just an addition to the noise contribution of other devices, and they must be spread out enough so that the origin of noise is not readily identifiable. This may prove tricky when there are such acoustic weaknesses in the ceiling as air grills and lighting fixtures. An implementation inside the technical floor void is also a possibility, and due to the high attenuation of the technical floor elements, it is especially hard to find by the average user.

Modern office buildings now heavily call for the use of the thermal inertia of the building. This has a significant consequence on the acoustics of the premises, as thermal engineers will often reject the presence of any ceiling under the upper floor due to its prevention of radiation and convection. However, experience has shown that it is possible, using either vertically hung baffles or horizontal panels under 60% of the upper floor surface, to achieve both the required thermal convection effects and the required acoustic absorption [23]. When using such a scheme, one will have to pay attention to the location of workstations with regards to exposed areas of ceiling in order to avoid undue noise propagation.

The importance of the various fittings should not be underestimated, both for their usefulness in noise reduction through either extra-absorptive surfaces or noise barrier effects, and for their visual influence on well-being [24].

One should remember that the design team is not the only one concerned regarding noise control. Here are a few hints:

- When introducing people to open space, have a short training dispensed to explain how one should not speak loudly and use the available private boxes for small meetings or lengthy or loud phone talks, and avoid playing music at one's workstation.
- Use office equipment that is as silent as possible.
- Locate noisy and frequently used common equipment, such as printers and photocopiers, in an enhanced acoustic absorption zone, and make sure one does not have to travel through the whole office floor to reach it.
- Especially avoid noisy equipment or people in a corner where the walls are not absorptively treated.
- Do not turn absorptive panels into message boards, as this will destroy their efficiency.
- Perform regular maintenance on the furniture (a chair with a lost rubber pad is not only uncomfortable to the user, but also noisy on a hard floor) as well as on the equipment (e.g., a badly maintained piece of equipment will squeak).
- Use absorptive office dividers (Remember: The closer to the source, the better for noise reduction).

7.5 MODELING

Modeling of a large open space can often prove tricky. To start with, a Sabine model will usually not perform well in an office environment due to the height being much smaller than the other dimensions, as well as (hopefully!) the ceiling being highly absorptive compared to other surfaces. On the other hand, a ray tracing model will usually experience trouble with diffraction (meaning that it will not properly take into account noise barrier effects), though some models now tackle this aspect [26].

Regarding restaurants and catering, where the situation is a bit simpler, some statistical models can help evaluate the risk of the cocktail party effect and point out how much absorptive surface is required as a minimum [27].

Unless the user has significant experience in this field, he had better rely on some established statistical models (e.g., [25]) and follow the relevant standards [2–5]).

7.6 MEASURING

Measuring is twofold: One wants to evaluate the acoustic characteristics of the space (e.g., reverberation time, sound level decay per doubling of distance, background noise level), and one also wants to know the sound pressure level at the workstation during operations. While there are standards regarding the former aspect (e.g., [28], with a sizable portion of them looking at the spatial sound level decay), the later aspect is harder to come by, as it involves the selection of representative workstations and a list of conditions to be reported; a tentative procedure is given in [4]. It must be pointed out that such measurements can be tricky, as most of the time there will be people attempting to prove that it is too noisy, while others want to keep quiet. This means that one cannot rely on automatic measurements, but needs to be close by and observe what is happening. A caricature of an example can be found in companies' catering areas: On seeing the microphone, some employees will lower their voice, but some others will adopt loud behavior; one of the tricks, then, is for the acoustician to arrive well before the first patrons and hide the measurement apparatus in, for example, a shelf of a flower pot (do not forget to warn the responsible restaurant staff beforehand and make sure the instrument only records sound levels for privacy protection purposes). The same pattern can be recognized in open-space offices. A simple trick to arrange with the manager is to find a workstation that will not be occupied on the measurement day and discreetly install the microphone on the back of the seat or on the shelf close to it so as to get all the usual noise contributions. Do be conscious that usually significant noise events will take place on arrival and departure of the staff, especially around lunchtime or tea break, and this may be of special importance if those events are staggered. Such events must be recorded. Be also on the lookout for such noisy behavior, as phoning on a mobile phone while walking in the aisles or close to the workstations, as this may point to either a lack of education or simply to a lack of private boxes for such purposes.

As the measurement goes on, do not hesitate to contact the staff and ask questions: They are the ones who spend their time in the premises, and they will be able to point out anything out of the ordinary. More often than not, they have some sound ideas about improvements, and more to the point, they can usually also explain most of the events to the acoustician (cf. Section 7.7.9) and help him rate the annoyance from those noise events.

Please note that some specific measurements (e.g., speech intelligibility between some workstations) may be needed [29] in some cases.

7.7 EXAMPLES

7.7.1 Director's Office in an Open Space (Failed)

Keeping with the spirit of transparency through open-plan offices, the management of a sizable society opted for a similar layout. Directors were located at the last floor of a newly refurbished building featuring plain plasterboard under the roof, which allowed for good indirect artificial lighting, and glazed partitions without doors, with a typical 12 m² per director. On being shown first the illustrations by the architect, then later the finished floor, they were enthusiastic.

However, within a couple of days of their moving in, they were disillusioned: Any phone talk, even at a low voice level, could be heard all over the place; there simply was no privacy at all.

When an acoustic engineer came, he could only measure at best a mere 18 dB attenuation between workstations. He then proceeded to explain to them that privacy required significant partitions, complete with doors, which implied tearing down the existing partitions.

Lesson Learned: The image of a project is not everything!

7.7.2 Director's Office in an Open Space (Passed)

Also keeping with the spirit of transparency through open-plan offices, the management of a large railway infrastructure company opted for a real open-space layout. But their design team did include an acoustician, who was duly briefed on the acoustic requirements of the various departments. In turn, he could formulate some proposals and discuss them with the interested parties, and each member of the design team could then adapt his prescriptions accordingly. Also, the directors were regularly submitted the project for comments and briefed about its limitations whenever of interest.

This resulted in a true open space with large distances (circa 10 m) between desks, where privacy boxes were also provided.

Lesson Learned: Discussion can often be the key for good decisions in a project.

7.7.3 Sustainable Development Office Tower

A large company decided to have an office tower built. In accordance with today's concerns, it was to be a sustainable development certified project. Accordingly, the acoustic engineer defined such targets as background noise levels and sound insulation inside the premises.

On completion of the project, two results were noteworthy: First, the office tower definitely was sustainable development certified; next, the users were clearly not happy with the acoustics.

Investigating this curious situation, while some resentment could be traced to the transition from partitioned offices to open-plan offices, it turned out that the background noise levels required by the sustainability standards were much too low to be of any use in masking noise in open spaces. Furthermore, the quest for natural lighting and fire regulations for high-rise buildings had prevented the application of acoustic absorption close to the workstations, which prevented the usual means of noise attenuation to be applied. In addition, it was also found that the number of privacy boxes provided at each floor had been badly underestimated, resulting in too many lengthy loud phone talks held from the desk position.

Lesson Learned: Building acoustics is one thing, and occupational noise is something else that must be investigated, in terms of both noise levels and significance of the noise.

7.7.4 Crisis Room

A crisis room is a meeting room where the analysis of a serious malfunction (e.g., a plane accident) is performed. Due to the likely presence of foreign or nonauthorized parties around, privacy with regards to the surrounding areas is essential. This means that suitable partitions and a set of doors must be built to ensure that people moving around do not manage to understand the talks held inside the room. It is assumed that due to the significant manning of workstations in the area, nobody will manage to stand close to either the doors or partitions.

7.7.5 Bank Management Floor

Privacy is essential in the premises of bank management. However, due to the number of people around, there might be a chance of somebody standing close to a partition or door and eavesdropping on what is being said inside meeting rooms. Therefore, one no longer considers mere speech privacy, but speech security for such spaces. This means that a higher sound insulation value is sought (e.g., 55 dB).

The trick was that in addition to rather efficient partitions, masking noise generators were installed in the ceiling close to the door and partitions. This resulted in an unfavorable signal-to-noise ratio for any would-be eavesdropper.

7.7.6 Dispatch Center

Dispatch centers provide the acoustician with a challenge: While they basically are open-space offices, the operators may wish for some discretion (so as to avoid disturbing their colleagues while on the phone or radio), yet they crave for whatever piece of information that can be heard from another colleague. A typical dispatch center will feature groups of desks treating either a defined geographical area or a specialty of interest. Incoming calls are taken by a frontal desk where a general practitioner will decide which specialty desk is best suited to deal with it. The specialty desk will then allocate the required means to deal with the request. In order to successfully answer, one nevertheless must make sure those means have not already been taken over by another desk.

This means that while the acoustician will try to reduce the noise levels through the use of absorptive materials in the dispatching room, efforts will be made to enhance speech intelligibility. This often results in a heavily treated open space with separate groups of specialty desks 6 m distant from each other. In addition, privacy boxes are also provided.

Examples of such dispatching centers include emergency coordination centers (typically made of firemen, doctors, paramedics, and police) in charge of evaluating the emergency, locating and sending the relevant vehicles and crews, and alerting the relevant response center (hospital, technical base), switch, and control towers.

7.7.7 Standard Offices

Standard offices are simply built in order to provide the users with some decent attenuation between workstations so as to avoid unnecessary annoyance, and are not meant to provide privacy. The usual construction calls for removable partitions installed between the technical floor and suspended ceiling.

Such a construction was used by the end user of an important industrial company where inevitably a few meetings would turn into a shouting match. During such a meeting, the acoustician and the architect were politely required to go to another office while the end user team sorted out their differences. While this would have been an acceptable solution for normal voice level, it was just insufficient for shouting, and both amused individuals were treated to a superb collection of epithets! Eventually, this was understood by the participants, and a grinning end user came asking whether privacy still was a valid concept for shouted voices!

7.7.8 Open-Plan Office

The acoustician was required to investigate the situation at a large research facility where some people were apparently complaining of noise issues. The typical layout featured two spaces of eight workstations 3 m distant from each other, with a half-height partitioned technical space housing the printers and plotters in between. The sound absorption under the ceiling was rather limited, while the ventilation noise was really low with 30 dB(A) in full operation. The ambient noise was a mere 45 dB(A).

Asked about the situation, one guy in the first space pointed out the lack of privacy, as he felt everybody could understand him during phone conversations, the annoyance from other people's occasional phone talk, and the unpredictable noise from the technical equipment that made him jump. He expressed deep regret at having lost his beloved former partitioned office. On the other side of the technical space the talk was different: Whenever something happened, everybody was notified at once by either raising one's voice or gesticulating, which made for real-time coordination. Regarding the technical equipment, everybody concurred that it was easy to hear whether the required plotting or printing had been initiated! They all expressed satisfaction at the new layout.

Lesson Learned: Open space can be accepted if cooperation is needed between members of a group.

7.7.9 Activities

On auditing an office facility where people were complaining of high noise levels, the acoustician was very surprised to find out that a cleaner would come around 10:00 a.m. and methodically sweep the room. When asked about the noise, the workers were adamant it did not bother them (though a 65 dB(A) equivalent noise level was measured at the workstations during the 15 min long operation). They pointed out that it was nice to chat with the cleaner and they could, when needed, ask him to put some special effort in cleaning specific parts of the room! Meanwhile, the sound level from another worker was considered a real problem, though it did not reach more than 45 dB(A).

Lesson Learned: Do not take anything for granted!

REFERENCES

1. AFNOR NF X35-102: *Conception ergonomique des espaces de travail en bureaux* (*Ergonomical design of office workplace*, in French), Paris, 1998.
2. AFNOR NF S31080: *Acoustique—Bureaux et espaces associés* (*Acoustics—Offices and associated spaces*, in French), Paris, 2006.
3. A.C.C. Warnock, Acoustical privacy in the landscaped office, *Journal of the Acoustical Society of America*, 53(6), 1535–1543 (1973).
4. *Acoustique des espaces ouverts* (*Acoustics of open plan offices*, in French), Standard Project NF S 31.199, St. Denis, 2013.
5. J. Bradley, B. Gover, *Criteria for acoustic comfort in open plan offices*, NRCC Publication Nr 47331, Ottawa, 2004.

6. V. Hongisto et al., Determination of acoustic conditions in open offices and suggestions for acoustic classification, presented at Proceedings of ICA 2007, Madrid, 2007.

7. Certivea, *Référentiel pour la qualité environnementale des bâtiments—Bâtiments tertiaires* (*Referential for the environmental quality of buildings—Office buildings*, in French), Paris, 2012.

8. BREEAM, *Environmental and sustainability standards*, 2008.

9. W. Passchier-Vermeer, W.F. Passchier, Noise exposure and public health, *Environment Health Perspectives*, 108(1), 123–131 (2000).

10. ISO 3352: *Acoustics—Noise assessment according to its influence on speech intelligibility*, Geneva, 1974.

11. M. Concha-Barrientos, D. Campbell-Lendrum, K. Steenland, *Occupational noise—Assessing the burden of disease from work related hearing impairment at national and local levels*, WHO, Geneva, 2004.

12. ASTM E1130-08: *Standard Test method for objective measurement of speech privacy in open plan spaces using articulation index*, 2008.

13. L.J. Loewen, P. Suedfeld, Cognitive and arousal effects of office masking noise, *Environment and Behavior*, 24(3), 381–395 (1992).

14. ASTM E1374: *Standard guide for open offices acoustics and applicable ASTM standards*, Philadelphia, 1993.

15. ASTM E1573-09: *Standard test method for evaluating masking sound in open offices using a weighted and one third octave band sound pressure levels*, 2009.

16. J. Veitch et al., *Masking speech in open plan offices with simulated ventilation noise: Noise levels and spectral composition effects on acoustic satisfaction*, Internal Report IRC-IR 846, Ottawa, 2002.

17. Bessiere, Commercial brochure, http://www.bessiere.pro/acoustique.php.

18. Makustik, Commercial brochure, http://www.akustik-raum.ch/en/acoustics-systems/products/acoustic-mineral-glass-mg/.

19. ISO 17624: *Acoustics—Guidelines for noise control in offices and workrooms by means of acoustical screens*, Geneva, 2004.

20. M. Asselineau, Noise levels inside open space facilities, case studies, presented at ICSV18 Proceedings, Rio, 2011.

21. J.A. Veitch, K.E. Charles, G.R. Newsham, C.J.G. Marquardt, J. Geerts, *Workstation characteristics and environmental satisfaction in open-plan offices: COPE field findings*, NRCC-47629, National Research Council Canada, 2004.

22. *BREEAM Europe commercial assessor manual*, SD5066A, Watford, UK, 2009.

23. Y. Le Muet et al., Thermally activated cooling technology (TABS) and high acoustic demand: Acoustic and thermal results from field measurements, presented at Internoise 2013 Proceedings, Innsbruck, 2013.

24. C.J.G. Marquardt, J.A. Veitch, K.E. Charles, *Environmental satisfaction with open-plan office furniture design and layout*, IRC Research Report RR-106, Ottawa, 2002.

25. FIOH, Room acoustic modeling of an open plan office, Helsinki, 2007, http://www.ttl.fi/openofficeacoustics.

26. P. Chevret, J. Chatillon, Implementation of diffraction in a ray-tracing model for the prediction of noise in open-plan offices, 132, 5, 3125–3137, 2012, Acoustical Society of America.

27. Peutz, Cantine, predictive assessment of noise levels in catering areas for schools and companies, Internal report, Paris, 1990.

28. ISO3382-3: *Acoustics—Measurement of room acoustic parameters—Open plan offices*, Geneva, 2012.

29. A. Ebissou, P. Chevret, E. Parizet, Objective and subjective assessment of disturbance by office noise—Relevance of the use of the speech transmission index, presented at Acoustics 2012 Proceedings, Nantes, 2012.

Chapter 8

Dwellings Hotels and Hospitals

8.1 INTRODUCTION

Dwellings hotels and hospitals acoustics mainly revolve around a common concern: limiting the amount of noise transmitted inside rooms and achieving a reasonable degree of privacy. The term *noise* may represent the noise in the next dwelling from, for example, discussion, a person walking, or user's equipment operating, as well as the noise from appliances and mechanical services (sanitary installations, HVAC, etc.).

Yet, regarding hospitals, there also is a need for the staff to be able to hear any suspicious noise that might spell alarm. In addition, there also is an emphasized need to prevent potential infection by means of room envelope coverings, as well as by mechanical services.

8.2 REQUIREMENTS

8.2.1 Regulations

8.2.1.1 Building Performance

Most countries now have regulations covering the field of noise control in dwellings [1–3]. Those regulations typically cover the following topics:

- Sound insulation with regards to the outside environment
- Sound insulation with regards to other spaces within the building (i.e., other dwellings, activity room, parking, etc.)
- Impact sound transmitted from other spaces
- Reverberation control
- Noise from mechanical equipment of the building

An overview of the acoustic targets required by a few regulations in hospitals has been given in a paper by Evans [4]; an overview of the acoustic targets required by some regulations in modern dwellings has been given in a paper by Rasmussen [5].

In addition, the noise radiated by the mechanical equipment in the environment will typically be subjected to the community noise control regulations.

In addition, one may notice that while stand-alone apartment buildings are quite common, one may also encounter dwellings in other kind of buildings, for example, in a school or in a facility, in order to house the staff. In addition to the usual dwelling acoustics regulations, some other requirements may be applicable too (e.g., in a discotheque or a cinema the dwelling will have to be considered as a third party and acoustically protected accordingly).

8.2.1.2 Occupant's Behavior

It must be stressed that the above-mentioned regulations pertain to the acoustic performances of the building housing the dwellings. It does not tackle the problem of noise emitted in the dwellings because of improper behavior, which is normally covered by community noise control regulations.

8.2.2 Standards

8.2.2.1 Building Performance

In addition to the regulations in force, there often are standards defining a somewhat higher acoustic quality. Such standards are meant to acknowledge the efforts made to produce a better-than-average building (and eventually justify a higher rent too!).

For example, in France there used to be the Qualitel [6] standard, which typically aimed for acoustic objectives better than the regulations by 3 dB. In addition, it also introduced some basic requirements regarding sound insulation between rooms of a same dwelling.

8.2.2.2 Occupant's Behavior

In some countries there may be a standard defining what the normal behavior of the occupant of a dwelling is supposed to be; for example, in France there is a national standard [7] titled *Normal Use Conditions of a Dwelling* stating that the noise levels in terms of L_{Aeq} should not exceed 65 dB(A) in a dwelling. The same standard also points out that people are required to put their slippers on and avoid undue activity noise.

8.2.3 Contractual Requirements

8.2.3.1 Building Performance

The end user of a project may wish for some specific acoustic target to be reached. For example, there might be a specific requirement on the sound insulation of the façade, for example, 5 dB or more over the value required by the regulations. There might also be a requirement on the noise of mechanical services or on the impact sound reduction. Last but not least, the end user may wish for a specific room to be thoroughly insulated in order to turn it into a studio or a rehearsal room.

8.2.3.2 Occupant's Behavior

It is not unusual for a landlord to issue a set of rules regarding the expected behavior of tenants. Among a variety of petty requirements (e.g., bicycles shall not be left by the entrance), there often are items regarding what is considered noisy behavior to be prevented (e.g., talking loudly in the public spaces of the building, listening to loud music, and being agitated after normal hours), as defined in the rules.

There also are some rules regarding the eventual work that is allowed inside a dwelling. For example, there often is a contractual requirement forbidding the owner or tenant to replace the floor covering by a material of his own choice due to the possible consequences regarding the transmission of impact sound in the building.

8.2.4 Rehabilitation

All the regulations above apply to new construction. What about rehabilitation projects? Well of course it usually is hard to achieve acoustic performances similar to those of a modern building in a structure that is more than 50 years old.

The usual basic target is to try to avoid degrading the acoustic performance of the existing building. This means that an acoustic diagnosis will have to be performed prior to any design work to find what those performances actually are. This will be a contractual reference point, as well as a basis for the design of the rehabilitation project.

8.3 ACOUSTIC TARGETS

8.3.1 New Dwelling Construction

The value of the acoustic targets must of course comply with the regulations in force. Those will typically cover the following items:

- Sound insulation with regards to the outside environment; the façade sound insulation value is typically set to limit the background noise level inside the living spaces (bedroom, living room) to the 30 to 35 dB(A) range.
- Sound insulation with regards to other spaces within the building (i.e., other dwellings, activity room, parking, etc.); the sound insulation value in terms of D_{nTw} is typically in the range 50 to 60 dB.
- Sound insulation within a flat or house usually is not a legal requirement, though it may be recommended by some high-quality standards (e.g., [6]).
- Impact sound transmitted from other spaces; the impact noise in terms of L'_{nTw} is typically in the range 50 to 60 dB.
- Reverberation control usually is not required within the dwelling proper. But several regulations require it for noise control inside common spaces, on the one hand, and for the comfort of hearing-impaired people, on the other hand (e.g., [8]); typically an equivalent absorptive area amounting to 25% of the floor area is required for corridors or waiting areas.
- Noise from mechanical equipment of the building; first, there is a distinction made between the common equipment of the building (e.g., boiler, lift) and the individual equipment of the dwelling (e.g., air conditioner). The former must not generate in the living spaces a sound level value greater than, typically, 30 dB(A), while the latter typically is allowed a higher value inside the dwelling where it is located. Next, there usually is a distinction made between permanently operating equipment (e.g., used air extraction) and random used equipment, which typically is permitted a 3 to 5 dB higher noise level.
- It is not unusual for some high standard dwellings to feature some unusual (as far as regulations are concerned) spaces and pieces of equipment. While home cinema facilities now appear, there also are music rooms, sport rooms, Jacuzzis, and even swimming pools (cf. Section 8.5.11) to contend with. Such spaces must be studied on a case-by-case basis, with required data to be provided by the client or the manufacturer (e.g., noise levels likely to be produced, periods of use, sound power level of the equipment of interest, velocity levels likely to be produced in a concrete slab similar to the floors of the project, etc.). Only then will the sound insulation, impact insulation, and sound power level targets be defined in order to comply with the applicable regulations (and possibly with the condominium rules in force too).

8.3.2 Hotels

While the acoustic problem of hotels is similar to the one of dwellings, there usually are some slight differences. To start with, a hotel will typically feature a dining room; it may even feature a ballroom. One also has to inquire with the operator about his intended schedule: while a small hotel usually has terminated most of its operations by 10:00 p.m., a luxury hotel can easily start processing a five-course dinner at 3:00 a.m. if the client wants it!

Some countries usually have a specific regulation covering the noise control of hotels (e.g., [9]). While the typical acoustic targets are inferior by 3 dB to that of a dwelling (e.g., a sound insulation value D_{nTw} of 50 dB between bedrooms [9], compared to 53 dB between regular dwellings [2]), one may also find some specific dispositions regarding the sound insulation of such noisy public spaces as the dining room or the ballroom, with the latter typically being subjected to the musical venues regulations (e.g., [10]). It is of primary importance to have the operator state his intentions for the various spaces (ballroom, sport room or swimming pool, bars and restaurants, etc.), in terms of both noise levels and periods of use (e.g., a ballroom operating for part of the guests while other guests attempt to sleep is a far different proposition from a ballroom rented by a group occupying the whole hotel). Section 8.5.12 illustrates what the stakes might be in some projects.

A distinction must be made between hotels and residences (as a reminder, a hotel provides fully furnished rooms with full maid service, while a residence usually has larger suites but basic maid service). The latter is typically considered dwellings, while the former is only subjected to the less stringent hotel regulations.

8.3.3 Hospitals

While the acoustic problem of hospitals is similar to the one of dwellings, there usually are some differences that eventually require a specific regulation [11]. To start with, the staff needs to hear any undue noise that may signal a problem. This means that the required sound insulation between rooms will often be much less than the one required in dwellings (e.g., a D_{nTw} of 45 dB). Next, there is a need for easy and efficient disinfection of all surfaces, which means that materials must be able to be thoroughly cleaned. Next, larger door and corridor dimensions must be taken into account to allow for the circulation of trolleys and wheelchairs.

Privacy must be provided to doctor's offices and examination rooms. This is valid for the space next to it as well as for the waiting area outside it. Section 8.5.8 gives a nice idea of consequences should privacy be omitted. How is one supposed to achieve privacy when doors are not that efficient and distances between speaker and eventual listener are rather short? Sound masking (cf. Section 3.10) may provide a rather simple answer under such circumstances: By increasing the background noise level on the receiving side (e.g., in the waiting room), one helps increase the privacy index and enhance privacy between emitting and receiving space without having to call for extra-absorptive material difficult to clean. Typical spaces to be outfitted are waiting rooms, corridors, patient rooms, and nursing areas.

There are also some rather noisy spaces (e.g., MNR room) that need to be thoroughly insulated.

8.3.4 Rehabilitation

To start with, an acoustic diagnosis must be performed prior to any attempt at designing the rehabilitation project. It is of primary importance for the design team to be properly coordinated, as clearly enough, once the structural engineer has performed his destructive testing, it will no longer be possible for the acoustician to try to measure in what is left of the rooms! Ideally, the acoustician should have been warned regarding the composition of the various floors of the building. While modern construction (i.e., 1960s and later) usually features rather regular structural elements in older European buildings, the ground floor would often be allocated to shops and workshops, the first and second floors would typically house the wealthiest tenants, while the floor under the roof would typically house the servants.

Inevitably, the acoustic performances reduce with height! Therefore, it is necessary to assess those performances in order to have a reference before any work is attempted, and also to be used as a basis for the new design. According to the available floor height and surface area, and budget too, objectives ranging from 48 to 55 dB in terms of D_{nTw} can be considered.

8.4 A FEW BASIC RULES

Typical heavy construction calls for 20 cm thick concrete slabs and 18 cm thick concrete walls. It is, of course, possible to use plasterboard walls (this will save weight, which usually is especially useful during rehabilitations), but one may kindly remember that apart from the acoustic issues, there also are some practical issues (such as the hanging of a portrait on the wall without penetrating through into the neighbors' space, and the security of the premises), so it is better not to use it for the perimeter of the dwelling.

When applying the finish of a wall, it must be stressed that glued plasterboard should not be allowed, as it will significantly degrade the acoustic performance of the structural wall. A regular plaster or concrete finish should be used instead.

Regarding impact noise control, it is possible to use a resilient floor covering. Now here is a simple experience one can do: Put a piece of carpet on a tiled floor, go in the room underneath, and ask a colleague to walk around with hard-soled shoes. You will be able to easily tell whether he is walking on the carpet or on the tiles. Now ask him to remove his shoes: You will probably not be able to distinguish between the two floor coverings. The lesson is as follows: Resilient floor coverings are nice as long as one deals with high-frequency impacts, but they are useless when it comes to low-frequency impacts. In such cases, a floating screed or even a floating slab is required. In both cases, it does take some extra space if it was not planned right from the beginning, and it may be prone to bad execution unless built by trained workers and supervised by the design team.

One should identify the spaces that need a heavy floating slab or even a box-in-box construction as early as possible in the design process, as such dispositions will impact the structural and architectural brief of the project.

When it comes to rehabilitation, some manufacturers have designed a rather nice constructive element made of a lost cast on which resilient pads are glued underneath and mineral wool are glued between the pads. When the structural capabilities of the building do not allow for extra concrete, two plywood sheets are screwed on the upper side of the casting. This constructive scheme can bring an impact reduction of up to 15 dB, while the sound reduction of the floor usually is improved by 3 to 5 dB.

Do be careful if changing the floor covering during a rehabilitation project: This usually is considered by the court as a degradation of the acoustic performance of the building.

In order to reduce the risk of annoyance between dwellings, care must be applied to ensure that potentially noisy areas such as kitchens and bathrooms are grouped together. It will not do to have the kitchen or bathroom of a dwelling right above a bedroom of another dwelling.

Incidentally, relocating the bathroom or kitchen over a living space during a rehabilitation project will also be considered by the court as a degradation of the acoustic performance of the building.

Do leave some space for the mechanical services. While it is sometimes difficult to squeeze in the various ducts and pipes in a rehabilitation project, there is no excuse for not allocating proper space for such services in a modern building. Remember, the more complicated the network (e.g., with numerous bends), the noisier it will be. Also, the smaller the section of the ducts, the higher the speed of the fluid and the noisier the result will be.

It is not uncommon, at least in Europe, to have commercial facilities located at the ground floor of dwellings. This supposes a few precautions in order to avoid the flats being subjected to noise and vibration annoyance from the commercial activities. To start with, the sound insulation between those types of spaces has to be greater than the average sound insulation between flats (e.g., in French regulations it has to be greater by 5 dB [2]). More to the point, extra precautions are required regarding impact noise control, as heavy carts may be rolled in the commercial spaces. And the hours of operation extend well out of the usual day (i.e., it is not uncommon for a commercial facility to get deliveries well before 5:00 a.m., while sorting and cleaning may occur until midnight). When dealing with such a project, it will be necessary first to state clearly what the restrictions to the operator are (e.g., limitations in the operating hours, admissible noise levels, means of delivery, etc.). As there usually are contractual requirements applying to the building constructor and other rules to the operator, one must make sure every contingency has been covered. A floating slab will be in order in the commercial premises: Will it be built by the constructor or added later by the operator (in which case the required space must have been provisioned in the ground slab for such a purpose)? Also, some extra equipment (e.g., freezers, coolers, generators) may be installed by the operator, and those must be subjected to noise level limits so as to comply with community noise requirements (e.g., [12]).

8.5 EXAMPLES

8.5.1 Dwellings

An old art deco building on a busy avenue in Paris was subjected to a major rehabilitation. The diagnosis showed that while the sound insulation D_{nTw} between floors was a comfortable 51 dB, it was no more than 36 dB at the last floor. More to the point, the structure was typical of the time, with a steel frame clad with stonework and floors made of a steel structure with a wooden floor on top and a plaster ceiling underneath.

The acoustic targets that would be required for a new construction were reached through a combination of resiliently suspended ceilings and plasterboard partitions. The main difficulty was to find space for the insertion of ducts and pipes, which was solved through the abandonment of floor space close to the stairwell.

8.5.2 Rehabilitation Turned Sour

The new owner of an old flat in France decided on a major rehabilitation. This included the replacement of the old carpet on a wooden floor by a new marble finished floor on a floating screed, with a structural concrete floor underneath. The kitchen was also moved and relocated closer to the dining room.

Right after commissioning, the neighbors complained about impact noise and quickly called for an expert to the court. It was found that the acoustic performance of the refurbished dwelling actually complied with the targets required for new construction. However, the expert stated that the impact sound was higher than before, and the relocation of the kitchen led to more noise transmitted than before to the neighbor's bedrooms.

As this was deemed to constitute a degradation of the initial acoustic performances, the owner was ordered to reconstruct his dwelling to the original scheme.

8.5.3 Noisy Behavior

A middle-aged lady came to the acoustician and explained that she wanted to seriously improve the sound insulation of her bedroom. After agreeing on a price for the acoustician's job, the basic principles of sound insulation were explained to her, and in addition, the need to prevent vibration propagation from a room was also explained. Eventually she was asked about the kind of noise she wanted to protect her bedroom from.

To the astonishment of the acoustician, she answered that she indeed wanted to prevent vibrations coming from the bedroom too, as noise was only part of the problem to her! Eventually she bluntly told the acoustician that she wanted to lead an active and colorful sexual life without the neighbors being able to know about it!

Lesson Learned: Always inquire about the eventual source of noise!

8.5.4 Home Cinema

The owners of a flat decided to treat themselves to a home cinema. They bought a package supposed to cover the installation of the sound and image system as well as the acoustic correction of the room in which it was installed.

Unfortunately, the said package certainly did not include any provisions regarding their sound insulation of the cinema room with regards to its environment. Faced with the reactions of irate neighbors, the owners eventually had to dismantle the whole installation and build a box-in-box room to cope with the high sound insulation (more than 75 dB) required; fortunately, the structural characteristics and dimensions of the building were compatible with such a construction.

8.5.5 Music Room

The owners of a flat decided their child should be provided with a room where he could play loud music. They quickly discovered that such an activity simply was not possible inside a normally sized dwelling, and eventually wisely opted for the renting of a specialized rehearsal space.

Lesson Learned: Practicing loud music simply is not possible in a flat.

8.5.6 Hotel

An old building whose origin dated back to the 1850s had been steadily expanded over the years, with extensions done as late as the 1920s. It was eventually acquired by a hotel company to be turned into a luxury hotel.

The diagnosis showed clearly that the constitution of the floors could greatly vary, not only from floor to floor, but also at the same level. First, the various compositions were identified and the relevant prescriptions were edited by the design team in order to satisfy the requirements. All separating partitions were made of thick plasterboard partitions. While the thickness of the concrete floor was kept to a minimum for structural reasons, it was sufficient enough to prevent flanking transmissions. Thick carpeting was used in most of the spaces. This eventually resulted in a hotel building complying with modern standards.

8.5.7 Hospital

When designing a brand new hospital in France, the design team was careful to locate the rooms by functional entities. One of the tricks was to have the patient rooms located in the back of the building, while the noisier, busier spaces, such as the hall, restaurant, and emergency areas, were at the front. Careful selection of separating floors and walls, together with judicious space planning, enabled the acoustic targets to be complied with.

8.5.8 Medical House

A medical house was installed in the lower floors of a combination commercial and housing building. A space planner was duly commissioned to fit the required offices and examination rooms inside the building. The doctors were appalled by the acoustic result: To start with the most obvious, the quest for space had resulted in the doors of two examination rooms being located in front of a structural column, which ruled out any stretcher case entering. Next, the space quest had also resulted in thinner than reasonable plasterboard partitions that did not provide the required sound insulation between rooms. Last, the doors were not sound-proof at all. A good illustration was given to the acoustician called to try to solve the mess by the speech therapist: He explained that he had to deal with a rather refractive boy, and on the first visit he eventually managed to make him speak. On the second visit, the boy stayed mute and eventually burst stating that on the previous visit he had been told that everything said would stay between them, yet his mother in the waiting room had heard everything!

8.5.9 Dwellings with a Supermarket Downstairs

An estate broker had a building erected with mainly dwellings. The ground floor and part of the basement were allocated for possible use by a commercial operator. They were eventually rented by a supermarket. The neighbors, especially people living on the first floor, were quick to complain about the noise: Cleaning occurred very late in the evening, while delivery could take place as early as 4:00 a.m. An expert to the court was mandated. He found that while the sound and impact insulation performances were up to the legal requirements, the noise generated by some activities was too high to comply with community noise regulations. The impact noise generated by forklifts was especially pointed out.

The operator was eventually required to build a floating slab; more to the point, he was also required to reduce his operating hours, with deliveries not allowed before 6:00 a.m.

This court action also found the builder partly responsible, as the commercial premises had clearly not been designed against such noise annoyance, and the tenant had not been warned about the potential limitations.

8.5.10 Dwellings with a Musical Venue Downstairs

In order to keep some activities downtown in the evening, a township decided that a building intended for dwellings would also feature a leisure place downstairs. The estate broker who was retained for this building operation decided on the ground floor allocated to a discotheque and first floor to lounges and offices, with the higher floors for dwellings. A project design was drafted by an architect and work started in earnest. However, there was a small point that had been overlooked: The authorities duly warned that the building permit was issued with a requirement for an acoustic study. The acoustician who eventually came was aghast: He was required to perform an acoustic study for a building permit, but the construction work was well underway. He nevertheless managed to issue a few warnings and prescriptions that were not heeded. Inevitably, the moment the facility opened there were numerous complaints. The expert to the court found that the doors between ground level and first floor were kept opened. More to the point, the sound system of the discotheque had been also installed in the lounges. Last, low-frequency sound was sent throughout the structure of the building due to loud bass music. After careful examination, the judge eventually ruled that the operation of such a facility was not realistic, as not only would one be subjected to the noise of activities inside, but also there was the noise from guest's cars outside, as well as the presence of people smoking outside and talking loudly. The facility eventually closed.

8.5.11 Dwelling with a Swimming Pool

While the construction of new dwellings was well under way and the relevant property put up for sale, the estate broker had the surprise to face a potential customer who declared himself quite interested in buying the whole last floor plus part of the floor underneath in order to have a sizable swimming pool added to his flat. After consultation, the design team managed to specify the required extra concrete thickness of the relevant floors and walls to be able to cope with the weight and keep the vibrations generated by the swimming pool pumps to a value that would prevent their being heard in other parts of the building. More to the point, extra noise control measures also had to be implemented to prevent the music in this area from being heard.

8.5.12 Hotel with a Large Ballroom

When a luxury hotel was constructed in Lyon, it featured a large ballroom over the main lobby and under the first level of rooms. As a sizable portion of the guests would be air-line personnel recuperating from a long flight while a wedding party might occur in the ballroom, the end user sternly reminded the design team of their duties and told them that should a noise problem occur because of noise transmitted from the ballroom to the guest rooms, the design team would be reputed to rent those rooms for life!

REFERENCES

1. A. Carvalho, J. Amorim, Acoustic regulations in European Union Countries, presented at Conference in Building Acoustics "Acoustic Performance of Medium Rise Timber Buildings" Proceedings, Dublin, 1998.
2. Arrêté du 30 juin 1999 relatif aux caractéristiques acoustiques des bâtiments d'habitation (Arrest pertaining to acoustic characteristics of dwellings), *Journal Officiel de la République Française*, 149, 10658–10660 (1999).
3. Arrêté du 30 mai 1996 relatif aux modalités de classement des infrastructures de transport terrestres et à l'isolement acoustique des bâtiments d'habitation dans les secteurs affectés par le bruit (Arrest pertaining to the classification methodology of transportation corridors and to the sound insulation of dwellings in zones concerned by noise, in French), *Journal Officiel de la République Française*, 49, 9694 (1996).
4. J. Evans, C. Himmel, Acoustical standards and criteria documentation of sustainability in hospital design and construction, presented at Acoustics 2012 Proceedings, Nantes, 2012.
5. B. Rasmussen, Sound insulation between dwellings—Requirements in building regulations in Europe, *Applied Acoustics*, 71, 373–385 (2010).
6. Qualitel, Référentiel 2008, Paris, 2008.
7. AFNOR P05-100: *Conditions d'usage normal d'un logement* (*Normal use conditions of a dwelling*, in French), Paris, 1991.
8. Arrêté du 1er août 2006 fixant les dispositions prises pour l'application des articles R. 111-19 à R. 111-19-3 et R. 111-19-6 du code de la construction et de l'habitation relatives à l'accessibilité aux personnes handicapées des établissements recevant du public et des installations ouvertes

au public lors de leur construction ou de leur création (Arrest dated August 1, 2006, defining prescriptions for accessibility by reduced mobility people to public receiving facilities, in French), *Journal Officiel de la République Française*, August 2006.

9. Arrêté du 25 avril 2003 relatif à la limitation du bruit dans les hôtels (Arrest pertaining to noise control in hôtels, in French), *Journal Officiel de la République Française*, 123, 9106 (2003).

10. Décret n°98-1143 du 15 décembre 1998 relatif aux prescriptions applicables aux établissements ou locaux recevant du public et diffusant à titre habituel de la musique amplifiée, à l'exclusion des salles dont l'activité est réservée à l'enseignement de la musique et de la danse (Decree about prescriptions applicable to public receiving facilities and rooms using amplified music with the exception of dance and music teaching reserved rooms, in French), *Journal Officiel de la République Française*, December 16, 1998.

11. Arrêté du 25 avril 2003 relatif à la limitation du bruit dans les établissements de santé (Arrest pertaining to noise control in health facilities, in French), *Journal Officiel de la République Française*, 123, 9104 (2003).

12. Décret no 2006-1099 du 31 août 2006 relatif à la lutte contre les bruits de voisinage et modifiant le code de la santé publique (dispositions réglementaires) (Decree pertaining to community noise control, in French), *Journal Officiel de la République Française*, September 1, 2006.

Chapter 9

Other Spaces Used by the Public

9.1 INTRODUCTION

As implied by their denomination, other spaces used by the public are public spaces that do not have the glamor of performance halls or the functionality of offices. Yet they serve important functions in everyday life, and their good acoustics may be a factor in their correct operation. This chapter will briefly examine their highlights.

9.2 ANALYSIS OF REQUIREMENTS

9.2.1 Foreword

There usually are a couple of intelligibility requirements to take into account: First, there must be a good intelligibility between the speaker and the listener of interest; yet there must not be annoyance because of this talk. Most of the acoustic problem is summed up in this dual question; all that remains is to define where the speaker–listener couple will be and where the potentially annoyed people will be located. It also hints at some future difficulties with regards to the noise generated by mechanical equipment: Its value must be low enough to allow speakers and listeners to properly communicate, yet it may be looked for as masking noise for the areas that are not concerned by this communication; therefore, it must be loud enough.

9.2.2 Applicable Standards and Regulations

The moment a space is accessible to the public there probably is a regulation somewhere that will define a few points, such as the number of emergency exits, and an acoustic requirement, such as an absorptive area. In France, this requirement [1] is applicable to all potential waiting areas, and their equivalent absorptive area must be at least 25% of the floor area.

In Europe there usually is a requirement based on a European standard regarding the emergency sound system whose common intelligibility scale (CIS) must be at least 0.7 [2].

While there usually is no specific standard for such lowly places, there are standards applying to the so-called associated spaces of offices. One can find some inspiration in there. Such standards will typically give some recommendations pertaining to the background noise levels and to the reverberation time (RT) or the minimum absorption needed in such spaces. Sometimes when the spaces are similar to those used in educational facilities, some basic indications regarding sound insulation, reverberation, and noise from mechanical services can be found in the relevant texts.

9.2.3 A Few Points Worth Considering

As low as they may appear, other spaces receiving the public serve a utilitarian aspect and must not be neglected. Just think for a second of the entrance hall: This will probably be the first visual and acoustic impression of an institution for a visitor. Can one really afford to have bad acoustics in there?

9.3 ACOUSTIC TARGETS

The acoustic objectives can be as follows:

* Background noise level in the 35 to 45 dB(A) range (the particular value can be adjusted according to the eventual need of its acting as masking noise)
* Reverberation time in the 1.0 to 1.5 s range (depending on the volume and aspect of the room) according to the purpose of the facility (i.e., a medium-sized speech-oriented space should not feature a reverberation time over the 1.2 s mark)
* Spatial sound level decay in the range 2 to 4 dB(A) per doubling of distance

9.4 A FEW BASIC RULES

One might care to remember that as a general rule, the problem is primarily concerned with intelligibility. This means one must find the background noise level and the reverberation time, as well as the distance involved between speaker and potential listener, that have to be identified. Next, one has to try to assess the spatial sound level decay for discretion purposes.

In addition, one also has to provide acoustic absorption in public receiving spaces and waiting areas as per the legal requirements in force (e.g., [1]).

9.5 A FEW TYPES OF PUBLIC SPACES

9.5.1 Entrance Halls

Entrance halls may just be that: the point of entry in a building where one just passes through. However, there often is a security checkpoint or simply a hostess who will have to take down the name of the visitor, inquire about his host, and give directions.

This means that some intelligibility is required at the desk (especially as both the hostess and the visitor may find themselves talking English while it is not their motherly tongue). Meanwhile, some discretion is also advised with regards to other people waiting or staff.

Proper attention must be brought to the hot air curtains should there be any such device in there (though in modern halls one prefers a correct air lock to such energy-consuming devices), as they usually are located quite close to the reception desk.

In addition, should the entrance door be located under a dwelling, care should be exercised to ensure that the noise of the door closing is not a source of nuisance in the dwelling.

9.5.2 Shops and Shopping Areas

Shopping areas usually are large places where various shops or stands are sheltered. The basic idea is for such places to have a minimum of discretion between stands. This usually is achieved through some sound attenuation between stands (e.g., by means of an absorptive treatment under the ceiling) and some background noise (which is generated by the background music as well as by the presence of the public and the noise from mechanical services). One also has to make sure that verbal communication will be correctly possible between the customer and the attendant. Last but not least, as the attendants will probably spend their workdays in the shopping area, one had better provide them with suitable acoustical conditions.

Some attention must be brought to the mechanical services (HVAC, escalators) to prevent their noise from being a hindrance.

Some shops are located at the ground floor of a building that may feature dwellings (a frequent pattern in Europe). While the above requirements still apply, one has to pay particular attention to community noise control. Besides the usual noise from activities in the premises, one has to pay particular attention to trolleys. In order to reduce the banging noise occurring on the floor, one should try to have as smooth a floor covering as possible. In addition, deliveries can occur very early in the morning, and this issue has to be addressed too. A floating floor can usually solve the impact problem, but it must be implemented prior to the shop opening due to the complications involved.

9.5.3 Exhibition Halls

Exhibition halls will typically shelter a large number of stands. The acoustic problem is similar to the one of shopping areas. The one difference is the frequent presence of sound systems on the stands.

Basically, the trick is to provide as much acoustic absorption as possible under the upper floor or ceiling. This is more easily said than done due to the numerous suspension points and fire appliances located there.

Some attention must be brought to the mechanical services (HVAC) to prevent their noise from being a hindrance. On the other hand, this noise can also be used as masking noise to ensure a minimum of discretion between exhibitions. Due to the heavy amount of lighting and large attendance, those systems are quite present, and while their masking noise can help, it must not turn into an inconvenience.

9.5.4 Museums and Galleries

Museums and galleries can have a significant attendance. More to the point, guided tours may be conducted there. In some museums certain exhibits feature small loudspeakers to provide locally vocal information.

The basic idea here is to enhance the spatial sound decay through a good absorptive treatment on the ceiling and part of the walls. Next, some noise control must be applied on the HVAC and lighting equipment.

As the exhibits usually are not as mobile as in exhibition halls, it is sometimes possible to use absorptive noise barriers. Those are particularly handy when it comes to limiting the propagation of the noise from the small loudspeakers. A few guidelines are given in standard ISO 17624 [3].

9.5.5 Airport and Station Lounges

Airport and station lounges are places where travelers can wait until their departure gate or platform is announced. In the meantime, they usually expect to find some peace before their journey.

To start with, this means that the noise coming from the environment (mainly vehicle noise, be it rail, plane, road, or even a mixture of them all [4]) must be reduced. This can be achieved through a good sound insulation of the envelope, including air lock-mounted doors, and completed when the situation permits with absorptive noise barriers close to the noise source (e.g., 1 m high absorptive barriers close to the track, as commonly practiced by several railway companies). More to the point, in the case of an enclosed station (e.g., an underground station or a large covered station), it can be interesting to provide absorption on the underside of the roof and the lateral walls [5, 9]. (Well, this case will not be found at airports unless one rebuilds a Templehof-like structure! But one may then consider the trick of inclined façades to avoid sending too much sound energy by reflection on the other façades around, as well as applying absorptive material on the blast deflectors where applicable.)

Finding peace also means that the PA announcements must be intelligible; this means that travelers must be in the direct field of the relevant loudspeaker. In addition, there is a need for privacy, as one does not want to be unduly disturbed by people talking at the next row of seats. For both those conditions a good spatial sound level decay (e.g., at least 3 dB(A) per doubling of distance) must be provided. In order to manage those conditions, a good absorptive treatment must be provided on the ceiling and also on the walls whenever suitable. A carpet as floor covering is quite advisable too.

9.5.6 Libraries

Good news: Libraries are places where people do not talk, at least not loudly. Now for the bad news: People talking in a low voice or the sound of a pencil dropping on the table can really be disturbing under such circumstances. It is not really good to attempt to provide masking noise under such circumstances, as it will often be perceived as annoying; 38 dB(A) is a maximum value for such purposes. This means that one must look for noise barrier effects (e.g., using the shelves of documents whenever possible) and provide an efficient absorptive treatment on the ceiling and also on part of the walls whenever possible.

One might care to note that libraries in schools and universities are subjected to the relevant regulations. By extension, the acoustic objectives required for such spaces can be based on the regulations pertaining to educational facilities.

9.5.7 Bars and Lounges

Bars and lounges are places where one may seek some nice time with friends and listen to music. This is a bit of a paradox, for good intelligibility would suppose a rather low background noise level, and due to the music, this is hardly the case. More to the point, one is seldom alone, and the noise from various discussions is heard everywhere.

One way to try to satisfy both needs is to implement quite a number of rather directive loudspeakers so as to ensure that all listeners are in the direct field of a loudspeaker. This enables the sound level value to be decently low. Next, an absorptive treatment on the ceiling and part of the walls is needed to increase the spatial sound level decay.

In addition, bars are liable to generate noise in the neighborhood due to the noise generated in the premises and transmitted through the floors and walls to other spaces of the building. Depending on the regulations in force, acoustic objectives may be specified as a limitation of the emergence inside the neighboring space (3 dB in each octave band from 125 to 4000 Hz in France [6]), or in terms of a maximum sound level in the bar (90 dB(A) in Belgium [7]), or a mixture of both (e.g., a maximal emergence of 5 dB(A) when the background noise level is under the 30 dB(A) mark, or 35 dB(A) in all other cases, in Luxemburg [8]).

9.5.8 Restaurants

Different kinds of places can be found under the name *restaurant*. The large self-service facilities have acoustic problems akin to those of open-plan offices and are accordingly treated in Chapter 7. School restaurants are treated under Chapter 11.

The main acoustic problem with restaurants is: How do we limit the noise inside the facility, for both customer comfort and neighbor quietness? The first thing is to provide some acoustic absorption in the dining room. This is performed using an absorptive treatment of the ceiling (typically using a ceiling with an absorption coefficient of at least 0.7); such a ceiling will reduce sound propagation through the space. If need be, it usually is possible to glue absorptive material under the upper floor or existing ceiling in order to save height.

One should beware that there might already be a plasterboard ceiling under the upper floor: Such a ceiling is probably there for sound insulation and fire safety purposes and must not be removed.

Do not forget that such facilities will usually operate late in the night, cleaning the dishes and preparing the dining room for the next day. Also, there may be deliveries quite early, at dawn. It means that noise control with regards to the neighbors is an important point to be tackled.

In addition, a regular restaurant features exhausts from the kitchen. While this is no real issue in a stand-alone construction, it can be quite a hindrance when the facility is located at the ground floor of a building, as a duct will have to go up to the roof and the fan extraction unit will often be found there, with its noise control problem (cf. Section 9.6.3).

9.5.9 Court of Justice

A court of justice features a large waiting area for the visitors and at least one courtroom where the judges, the interested parties, and the audience will convene. In addition, there are offices, holding cells, and various support rooms.

There are quite a few acoustic aspects to take into account: To start with, one must ensure not merely privacy between some of those spaces and their surroundings, but actual security of speech (e.g., it will not do if the prosecutor located in an office can understand all the discussions of the defense located next door; cf. Section 9.6.9). This means that adequate partitions (featuring a sound reduction index of at least 65 dB) must be chosen, and attention must be paid to possible flanking transmission by the façade and adjoining partitions, as well as parasite transmission by means of the ductwork. More to the point, doors must be chosen with care regarding both their performance and their location (e.g., it will not do to have a door located close to a waiting area, and one had better forget about direct communication between offices); air locks should be considered whenever privacy is an issue (Remember: A high-performance single door may do the job on delivery, but it will not last).

Next, one has to provide good speech intelligibility in the courtroom (taking into account that some people will mumble at the stand, and will not be trained on the use of a microphone). This means that echoes and untimely acoustic reflections must be prevented and the background noise level kept to a minimum. The problem is similar to the one encountered in conference facilities. When dealing with old construction, one has first to perform an acoustic diagnosis and confer with the historical building architect to see whether some part of the inner envelope can be absorptively treated. It may also be possible to heighten the floor level in order to reduce the volume and the distance from speaker to ceiling. Another possible trick may be to lower the ceiling (cf. Section 9.6.10).

With new construction things can be a bit simpler, especially when studied from the beginning. One has to make sure that the dimensions of the main spaces (court of justice rooms and main hall) have been properly taken into account, including the relevant space for the ducting network and its silencers, as well as the proper space for floor-ceiling assembly and partitions. Acoustic absorption must be provided and a suitable sound system specified (Remember: It must be directive, and loudspeakers must face absorptively treated areas).

9.5.10 Sport Hall

A sport facility may range from a rather tiny room where a dozen people will exercise to a much larger space capable of handling two basketball teams and perhaps even their spectators too. Let's sort it out:

- Facilities meant to have sport teams and their public are usually also concerned with other shows; as such, they are tackled in Chapter 15 (multipurpose facilities).
- Facilities for educational purposes are covered in Section 11.5.7. This subchapter will deal with the kind of facility as a gymnasium or a swimming pool, either as a stand-alone construction or as part of another facility (e.g., in an office building or a dwelling).

To start with, there is a serious impact and sound insulation problem to be solved: Unless the space under consideration is a stand-alone construction, impacts from people jumping, balls thrown on the floor and walls (or even ceiling, for that matter), as well as running or weight-handling machines, will be transmitted to the structure of the building and radiated in sensible spaces of the building. Next, physical exercises are seldom performed in silence, and there often is some rather loud music to encourage performers. This means that when a sport room is not located in a stand-alone building, a floating slab on a plain thick concrete floor will be needed. More to the point, a box-in-box construction will probably be required to comply with the community noise applicable to musical noise or amplified music (e.g., [6]).

Sound insulation with regards to the outside environment will also have to be taken care of in order to avoid trouble with the neighborhood; just as a reminder, compliance with the relevant noise control regulations (e.g., [6, 10] in France) is not an option, but definitely compulsory! Of course, those requirements must be satisfied under normal ventilation conditions (which means that keeping the door or window open will not do!).

Next, proper absorptive treatment will have to be applied on part of the walls (two adjacent walls so as to prevent significant resonances) and the ceiling or roof underface. This is required to comply with the usual standard requirements (e.g., [11, 12]) regarding reverberation time, and also to enable the leader to deliver spoken instructions. One may care to note that shock-resistant absorptive materials will probably be needed (e.g., [13, 14]).

Last, the background noise levels inside the room or hall must be kept within reasonable values, for example, 45 dB(A). This means that industrial-type heating fans will not be acceptable.

9.6 EXAMPLES

9.6.1 A Small Bar

A small bar was opened in the end of the 1990s in Paris under the name Les Furieux ("The Furious Ones"). While he wanted his clients to enjoy rock music, the owner did not want to infuriate his neighbors. More to the point, he also wanted his clients to be able to enjoy a chat without having to unduly raise their voices. Last but not least, he wanted to keep the aspect of the beautiful stone walls and wooden beams inside the bar.

A small acoustic test was performed using a set of loudspeakers and a CD player; the test samples were made out of the owner's personal selection. As a safety, measurements were performed at the first floor level where the office of the owner was located: If the criteria were met there, they would be satisfied at the second floor, where the closest neighbors were. It turned out that in order to keep a decent quality without risking annoying the neighbors, the sound level of the music should not, in terms of equivalent sound level, be higher than 65 dB(A).

After some meditation on his concept, the owner eventually opted for a scheme where there were 64 small loudspeakers (for a 120 m² facility), so that clients are always close to the direct field of a loudspeaker. The concept has proven popular with the clients, and the neighbors have declared themselves especially happy, as they had previous bad experience of other operators for a bar there.

9.6.2 A Larger Bar

A famous musical bar in Paris underwent a complete reconstruction. The real acoustic challenge was the presence of a dwelling next door (to the extent that the bedroom was located on the other side of the wall).

As the new owner wanted freedom of operation in his bar, A heavy construction was devised. The envelope of the bar was built using a concrete structure with the foundations decoupled from those of the dwelling. Next, the bar itself was built as a box in a box. This scheme worked well enough, so that the owner could operate at the maximum permissible level (i.e., 105 dB(A) over the most noisy 10 min according to the French regulations), while the neighbor enjoyed his quietness with a 22 dB(A) background noise level.

9.6.3 A Restaurant

A restaurant was outfitted at the ground floor of a Parisian building dating back to the early 1970s. The inhabitants complained of the noise generated by its operations. More to the point, the customers were not happy with the noise levels during their meal. Eventually, an acoustician was hired to investigate the matter.

It turned out that the noise levels inside the dining room were rather high due to the high-density seating. In a feeble attempt to have some privacy, music was generated at too high a level. More to the point, in order to keep a rather flush ceiling, the loudspeakers had been incorporated in the supposedly fireproof and sound-insulating ceiling, and the back of the loudspeakers radiated under the upper floor. This was eventually solved through a complete reconstruction of the ceiling with resiliently suspended plasterboard for sound and fire insulation purposes under the upper floor, together with an absorptive decorative ceiling underneath that incorporated the loudspeakers that were laid with a higher density and a much smaller sound power level.

While this solved the dining room problem, much remained at stake: Deliveries were performed at dawn, and the inhabitants were especially wary of the rumble from beer barrels rolling on the incline to the basement. New methods, including the use of trolleys on tires, had to be implemented. The noise from the dish cleaning station was an issue in terms of both community noise control and occupational noise control. It was solved through the application of washable absorptive acoustic tiles, while the machine itself was treated to a new enclosure. The noise from the kitchen was actually satisfactory with the exception of a couple of issues: To start with, the staff would bring the large piles of cleaned dishes on the metal shelves that were structurally connected to the walls, and resilient material had to be applied. Next, the exhaust fan initially had to be installed in the kitchen, but the neighbor at the upper floor complained of its noise, and so it had been moved to the roof—right over the last flat that had been acquired by a pensioner who had looked for such a location so as to avoid noise from above! On looking more closely, the acoustician found that the resilient suspension of the fan was not working properly, but in addition, the staff did not bother with adjusting the speed of the fan. When it was cut close to midnight, the creaking sound of the decompressing metal duct could be heard all over the building. This was solved using an automatically adjustable variable-speed fan unit. Last, there was the issue of the trapdoor from the kitchen to the cellar that had to be closed by the staff on leaving the premises: It was banged shut around 2:00 a.m. to the anger of the neighbors. This was solved through specific instructions and a door brake.

Lesson Learned: Education of the involved staff is a must. In addition, many small noise control measures have to be taken into account to ensure a proper result.

9.6.4 A Museum

In Paris the Palais de Tokyo was built as the Japanese pavilion for the 1936 universal exposition. True to the architectural cannons of the day, it featured marble floors, stone walls, and plaster ceilings. The reverberation time in the empty exhibition rooms was over the 7 s mark.

When a project to turn this building into a museum of image was launched, the first question for an acoustician was: How do we achieve a minimum of sound attenuation between spaces of the future museum? While the same concern was shared by the end user, this was apparently not the case with the architect, who kept repeating that the acoustics of the place would be suitable to his architectural project. When no acoustic report was forwarded at the concept development stage, the end user was not happy; when no report was submitted at the design development stage, he was downright afraid and hired an acoustician to look over the project. There followed quite a lot of heated exchanges between the interested parties, with the architect accusing the acoustician of ruining his project with the proposed application of stupid absorptive surfaces or the insertion of doors at the entrance of some spaces. Even when the absorptive surfaces were classified as furniture, the architect would still object. Eventually, the project was dropped.

In a similar but smaller project, an interior architect wisely decided that acoustics was part of the issue; due to the architectural and technical constraints, the absorptive treatment was affixed on mobile panels that were discussed at length between the architect, the acoustician, the museum specialists, and the safety engineer. In addition to those panels, highly directive low-power loudspeakers were efficiently used.

In a more recent rehabilitation project a famous architect labeled the acoustician an "architectural terrorist" and refused any cooperation. It took the end user and the safety engineer to force the hand of the architect.

Lesson Learned: Acoustic treatment has to be discussed to get the best compromise between architectural and acoustical needs (sorry, but no compromise can be acceptable when dealing with safety).

9.6.5 An Exhibition Hall

The exhibition hall of a small town was initially made of a metal structure supporting metal panel walls and a metal roof. This was cheap construction, and it suffered from a few obvious defects: It was stifling hot in the summertime and it cost a small fortune to heat during the wintertime; this meant that in order to operate comfortably and economically, the operating periods were reduced to spring and autumn. More to the point, it was quite noisy due to the intrusion of road traffic noise into the hall, and also due to the lack of acoustic absorption inside. Last but not least, the neighbors complained every time when there was some event held inside due to the noise radiated by the structure, as the sound reduction index of the walls was estimated to be a mere 15 dB.

A renovation of the facility was eventually carried out. While the main structure was kept, the walls were modified using a thermoacoustic panel made of a thick mineral wool and a plain metal plate supplemented by an absorptive material and a perforated plate (with a sound reduction index over the 30 dB mark). The roof received a new waterproof coating and thermal insulation, and also got an absorptive material under it too. Doors were mounted air-lock style so as to enable the use of ordinary doors. Last, an efficient air handling unit (AHU) was installed on the roof of the facility. This renovation solved both the internal acoustics problems and the community noise problems for exhibitions.

9.6.6 An Entrance Hall

The office tower of a large company featured a sizable entrance hall. The end user listed the points he wanted to be taken into account: The receptionists should be able to understand the visitors, and the waiting area should be quiet enough. This meant that acoustic absorption had to be brought into the space; after fruitful discussions with the architect, it was decided to use a perforated plasterboard ceiling on most of the surface, with some mural complements using fiber panels close to the receptionists' and the waiting area. This was deemed successful by the receptionists and the visitors alike.

In another operation the client followed the advice of his architect, who kept pointing out that the hall must be majestic and, as such, should not be polluted by acoustic absorption. Within 2 months of operation, visitors were irritated at having to repeat their requests to the receptionists and the staff were fed up with working in such difficult conditions. A small "accident" conveniently destroyed the receptionists' desk, and this was taken as an opportunity to relocate it with suitable absorptive treatment applied on furniture elements. This solved the problem for the receptionists, and later on the waiting area for the visitors was similarly treated.

9.6.7 An Airport Hall

The airport building of Kansai International Airport, Osaka, Japan, was designed by Renzo Piano Building Workshop. It is close to 1500 m long and is erected on an artificial island.

In order to provide passengers and staff with satisfying acoustic conditions, it was of course first necessary to provide a good sound insulation with regards to the exterior due to the amount of transportation noise around the building: Apart from planes, there also were road vehicles and trains. This was achieved with façades and roofing of suitable acoustic performance. By the way, one of the difficulties was to achieve a reasonable sound reduction index value while avoiding too much weight increase, as in this seismic area one has to think of the consequences of an earthquake. Next, it was necessary to control the noise from mechanical services. This was achieved through a careful selection of the equipment regarding their emitted sound power level. A peculiar feature of the building is the concept of an open air duct used in the main hall. There are air outlets forcing air up onto an opened inverted channel from which it drops back onto the zones to be thermally treated. As those channels would cover a significant part of the underside of the roof, it was necessary to find a material that was both acoustically absorptive and aerodynamically suitable. The final result has proven satisfactory to both the passengers and staff.

9.6.8 A Library

In the frame of a large urban renewal project a new library was built in Strasbourg, France. The project used an old multistory shed of historical importance that was linked to a new concrete construction.

The sound insulation with regards to the exterior (where there is some road traffic noise and occasional boat noise too) was achieved using glazed façades of suitable acoustic performance. The point of friction with the architect came when dealing with the internal acoustics: The architectural master idea was to keep the upper concrete floor bare; no acoustic treatment was allowed on the grounds that the books would provide all the absorption needed. The ventilation of the premises was performed using round ducts of very clean lines, and no silencer was allowed on the louvers. Last, the floor covering was resin, with the signage embedded into it.

When commissioned, it turned out that the library, as fully fitted with books and shelves, met the acoustic objectives. However, the acoustician notes that the legal objectives for a library in France are only concerned with reverberation time and do not consider the notion of spatial sound level decay. Nevertheless, owing to the fact that the Alsacians are very disciplined people, noise never was a problem in this library.

9.6.9 A New Court of Justice

When a new court of justice was decided upon in a suburb of Paris, the various interested parties wanted to get away from the stereotypes of old cramped dark buildings. Consequently, an executive-looking building was eventually designed and erected. Trouble started right at the beginning of the project when the design team was given the job: When asked about a possible reduction of his fees as part of the effort to reduce the global cost of the project, the architect simply removed the acoustician from his team of engineers! On examining the preliminary design of the building, the technical controller inquired about the acoustics and was told it had been taken care of. When the design development was over and no acoustic targets had been forwarded, he knew there was a serious problem. By the time the end user reacted, the contractor had taken over and eagerly built what was specified. In order to keep a nice-looking aspect for the glazed façade as seen from the environment, the architect had specified thin partitions that proved insufficient to allow for privacy between offices. More to the point, the acoustician was refused the erection of a proof cell on the ground of economy.

Lesson Learned: A technical controller is always an added safety in a project, and a proof cell is the last chance to invalidate a serious design flaw.

9.6.10 An Old Court of Justice

When the renovation of an old French court of justice facility was decided upon, the end user was left with serious interrogations regarding the main court of justice hall. Its acoustics had been known to be bad for ages, to the extent that in the 19th century it had been turned 180° from the previous arrangement and its floor had been heightened to try to reduce its volume.

An acoustic diagnosis was carried out. Its showed that the RT was rather high at 1.5 s, and the sound system currently in use was not much good, as there were too many emission points that merely contributed to the background noise level. In addition, the windows featured simple glass that did not do much for thermal and acoustic insulation. Due to the presence of a painted ceiling, it was deemed impossible to do any work on this part.

The propositions from the diagnosis were multiple. To start with, it was recommended to provide better sound insulation to the windows (taking also the opportunity to provide better thermal insulation) and to reduce the noise from the ventilation through the implementation of a new ducting system complete with silencers and low air speeds. Next, it was proposed to reduce the acoustic height of the room by suspending Plexiglas panels under the historical ceiling. Last, a discussion with the historic building architect showed that the inclusion of absorptive materials behind the fabrics covering parts of the walls was a distinct possibility. More to the point, a new sound system with directive microphones and loudspeakers was specified.

Lesson Learned: There may be quite a few possibilities to improve the acoustic performance of existing buildings, but they have to be discussed by all members of the design team and the end user to make sure that they are fully compatible with each other's needs.

REFERENCES

1. Arrêté du 1er août 2006 fixant les dispositions prises pour l'application des articles R. 111-19 à R. 111-19-3 et R. 111-19-6 du code de la construction et de l'habitation relatives à l'accessibilité aux personnes handicapées des établissements recevant du public et des installations ouvertes au public lors de leur construction ou de leur création (Arrest dated August 1, 2006, defining prescriptions for accessibility by reduced mobility people to public receiving facilities, in French), *Journal Officiel de la République Française*, August 2006.
2. EN 60849: *Sound systems for emergency purposes,* Brussels, August 1998.
3. EN ISO 17624: *Acoustics—Guidelines for noise control in offices and workrooms by means of acoustical screens*, 2004.
4. Peutz, Kansai International Airport acoustic design development report, Paris, 1989.
5. Peutz Daidalos, Antwerpen Centraal acoustic design development report, Leuven, 2007.
6. *Décret 98-1143 du 15 décembre 1998 relatif aux prescriptions applicables auxs établissements ou locaux diffusant à titre habituel de la musique amplifiée (French decree on spaces using amplified music on a regular basis, in French), Journal Officiel de la République Française, 18957 (1998).*
7. Arrêté royal du 24 février 1977 fixant les normes acoustiques pour la musique dans les établissements publics et privés (Royal arrest defining acoustic norms in public and private facilities, in French), (Moniteur Belge), April 26, 1977.
8. *Règlement grand ducal du 16 novembre 1978 concernant les niveaux acoustiques pour la musique à l'intérieur des établissements et dans leur voisinage* (Earldom regulation concerning music inside facilities and in their surroundings, in French), *Parlementaire n° 2213, 1977–1978.*
9. J. Kang, *Acoustics of long spaces: Theory and design guidance*, Thomas Telford, London, 2002.
10. Décret no 2006-1099 du 31 août 2006 relatif à la lutte contre les bruits de voisinage et modifiant le code de la santé publique (dispositions réglementaires) (Decree pertaining to community noise control, in French), *Journal Officiel de la République Française*, September 1, 2006.
11. *Directives et recommandations (guide technique) pour l'aménagement d'installations sportives (Directives and recommendations (technical guide) for the fitting out of sport facilities, in French),* Service de l'Education Physique et du Sport, Canton de Vaud, Lausanne, 2012.
12. *Salles sportives—Acoustique (Sport halls—Acoustics, in French)*, Afnor, St. Denis, 1992.
13. Rockwool, Boxer ceiling, Documentation, 2012.
14. Eurocoustic, Acoustichoc ceiling, Documentation, 2012.

Chapter 10

Production Facilities and Workshops

10.1 INTRODUCTION

Production facilities and workshops are places where artisanal or industrial tasks are carried out. Some of those places are particularly busy and may feature numerous noisy pieces of equipment. As the purpose of this book is architectural acoustics, some aspects will be left out in this chapter.

As a reminder, there are three main aspects, as identified by the EU directives [1]:

- Noise generation by pieces of equipment
- Noise propagation inside the premises
- Noise reception at the workstation and noise exposure of the personnel

The former is reputed to be the problem of the equipment manufacturer, who must assess the sound power level of his equipment or its sound level at the workstation. The latter is the responsibility of the operator of the facility, who must protect his workers from undue noise exposure through the relevant management of spent time and appropriate noise control procedures (which inevitably entails a work analysis as described in ISO 9612 [2]). Noise levels limits are stated in national regulations (e.g., [3–7]). In between, noise propagation is the problem and the responsibility of the builder/fitter of the facility, and it will be the main topic of this chapter.

The purpose of this chapter is not to cover all aspects of noise control in such spaces. One will look at the implications on building acoustics. The reader interested in noise control strategies can refer to references [8, 9].

10.2 REQUIREMENTS

Let us consider a typical workshop. There is production equipment together with handling equipment to bring in the raw material and carry out the products. In addition, there usually is heavy air handling equipment, both for the comfort of the workers and for safety purposes.

When a specific worker is in such a space, he is subjected to the noise of his own activities and also to the various noise contributions coming from all the other noise sources inside the workshop. This means that the subject of noise propagation will be of primary importance.

Noise can be propagated directly (along the line of sight) from the noise source of interest (e.g., another production equipment or an air handling system), but it can also be propagated by acoustic reflections of the floor (which often happens to be painted concrete in such facilities) or on the walls and roof, as well as any acoustically reflective surface, such as a metal closet.

How does one describe the acoustics of such a place? Keeping in mind that first there will be a need for predictive computation (e.g., to check whether the new layout of equipment or the new architectural treatment of the workshop will actually constitute an improvement over the present situation, but also to assess whether the compliance with the acoustic requirements is met). The first obvious possibility in an enclosed space is the reverberation time (RT). However, looking at a typical workshop geometry, it turns out that the volume is rather flat (i.e., the height is small compared to the length or the breadth), and the RT will not account properly for additional absorption in an empty workshop. More to the point, in a fully fitted workshop, the volume fittings (e.g., pieces of equipment) will lower the RT value without the workers feeling better about it (cf. Section 10.7.9). Another way is to use the amplification at a given distance (typically 10 m) from a reference omnidirectional source, but it supposes that one does have a 10 m line of sight available, together with at least 3 m from the walls to the source or the receiving point, as well as a reference omnidirectional source. More to the point, this may be indicative of the acoustics in a given zone of the workshop. One can also use the spatial sound level decay as defined in ISO 14257 [10] using an omnidirectional source. While this takes into account a broader zone (e.g., 3 to 24 m from the source), it still requires an 8 m minimum line of sight available, together with at least 3 m from the walls to the source or the measuring line, as well as an omnidirectional source. Incidentally, one should note that corrections may apply to the sound source when it is not really omnidirectional or when the floor is not reflective [11].

In 1990 the French Ministry of Labour published an arrest [12] defining spatial sound level decay targets. Two cases were considered (empty or fitted workshop), and a typical rate of 3 dB(A) per doubling of distance was retained.

10.3 ACOUSTIC TARGETS

The actual ubiquitous requirement regard the daily noise exposure level of workers, which should not exceed a value defined in the relevant legal texts.

In addition, other legal requirements may apply. To start with, there will probably be a requirement regarding the amount of noise radiated to the outside environment (typically originating from the noise of equipment outside of the premises, as well as from the noise inside the premises transmitted through the walls and roofs of the facility). There may also be a requirement regarding the amount of absorptive treatment applied inside the premises (e.g., [12]), whose efficiency is assessed through the measurement of the spatial sound decay curve [14].

10.4 A FEW BASIC RULES

Should the workshop be delivered as a blank space (i.e., a space whose fitting out will be left to the end user), one had better prepare for the worst and apply absorptive treatment all over the place. But when the user has defined his layout, specific precautions may be elaborated, such as absorptive noise barriers around either noisy pieces of equipment or sensitive workstations. Practical recommendations regarding the use of noise barriers inside an enclosed space can be found in standard ISO 17624 [13]. When elaborating such schemes, one must keep in mind that other requirements may interfere: ease of access (for both operational and safety purposes), fire rating, ventilation, and lighting.

Do beware of the client merely answering "no need to consider this space, as it is not noisy." More often than not, "not noisy" means "less noisy than elsewhere"! Also, beware of the remark "there is nobody in there," as it may be true under circumstances defined as normal, but it will probably be false during a significant portion of the day (the reader may care to have a look at Section 10.7.2).

An apparently nice way to control noise emission is the use of enclosures—nice for the acoustician, that is. However, there are quite a few limitations to be conscious of: To start with, the process under the enclosure will receive materials and produce items; this means that apertures have to be provided. Next, ventilation (both supply and exhaust) also has to be cared for. Last, maintenance has to be easily practiced; this is of primary importance during production where a stoppage under the enclosure must be dealt with quickly. If the acoustician does not take into account the provision for correctly located access doors, it is quite certain that the maintenance staff will see to it—without further thoughts about acoustics. Incidentally, one looks for doors, not for removable panels: Experience shows that removable panels are dismantled for the duration of a specific maintenance work; often they are not mounted back after that work (supposedly to be able to perform faster adjustments), and gradually they are forgotten.

Controlling propagation throughout the work space will entail an absorptive treatment of the ceiling or the underface of the roof. When looking from the noise source (e.g., a piece of equipment), this is the part that is most visible (and exposed to sound radiation from the various pieces of equipment), as illustrated in Figure 10.1. What about the zenithal light brought in by a shed-type roof? Well, one can easily solve that using absorptive baffles, with a rule of thumb that there will be 1 m² of baffle per 1 m² of floor. Please note that such a treatment needs some height so if retrofitted, one will have to take care of fire detection and sprinklers, as well as the crane overhead (when things get too constricted, eventually one is left with the old trick of absorptive panels flat under the upper floor or roof). In addition, it will probably be useful to try to have an absorptive treatment (e.g., wood fiber panels) on part of the walls.

Noise barriers may be used to try to enhance further the spatial sound level decay. But there are a few limitations: To start with, due to the proximity of many reflective surfaces (e.g., machinery, walls, shelves), they will not be as efficient as the outdoors. Next, they do need to be absorptive. Last but not least, they must not be a hindrance for safety or operation.

Why bother with such complications as enclosures, acoustic absorption, and noise barriers, when one can simply ask workers to put on their earplugs? A natural answer can be found in the question: Does one seriously believe that by asking (or even ordering, for that matter), earplugs will be used? At any rate, the European directive [1] does promote collective protection over individual protection. More to the point, earplugs can be unreliable [15].

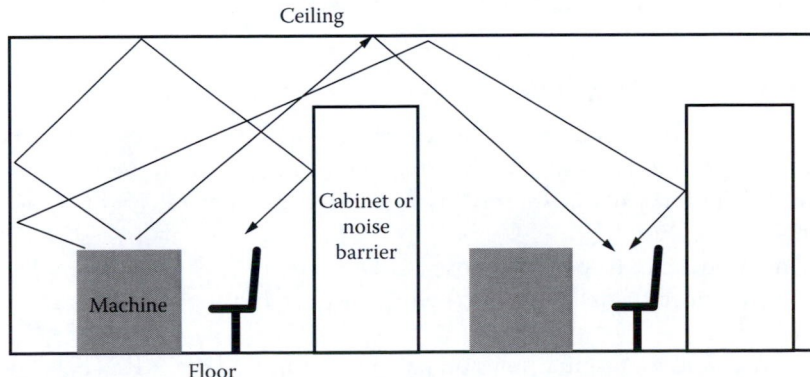

Figure 10.1 Effect of reflective surfaces (e.g., ceiling, walls, equipment) on the propagation of sound between workstations.

10.5 MODELING

Modeling of a production hall can be quite tempting for two purposes:

- Assessing the efficiency of the absorptive treatment (e.g., by computing the spatial sound level decay in the empty hall)
- Assessing the noise levels in the hall due to the operation of various pieces of equipment (this will help point out the main noise contributors, as well as pave the way for noise reduction solutions)

While it does not seem too complicated a prospect when dealing with the empty hall (whose shape usually is rather regular), things get pretty complicated with a fully fitted hall due to the numerous fittings to be found. Computation models and strategies can be found in [14].

10.5.1 Sabine Modeling

Sabine modeling is a very simple way to model a hall. However, Sabine's theory supposes a volume with a rather diffuse sound field and similar dimensions in the three directions. More to the point, it does not take into account a significant difference of treatment of a given surface (e.g., the ceiling) compared to other surfaces.

When looking at a real hall, the volume is usually considered flat, as the height is much smaller than the other dimensions. More to the point, usually there is only one surface treated to acoustic absorption, which is the ceiling.

The sound level L_p at a distance d from a sound source of sound power level L_w and directivity Q can be estimated using the formula

$$L_p = L_w + 10 \log(Q/4 \pi r^2 + 4/R)$$

where $R = A \, S/(S - A)$, with A equivalent to the absorption area of the hall and S the total area of the interior envelope of the hall.

Sabine modeling will overestimate the efficiency of the absorptive treatment. When performing a round-robin test on various prediction models in 1988 in an empty or moderately fitted hall, it was found that on a 30 m long, 3.8 m high volume, the predicted noise level at 20 m from the source could be underestimated by as much as 5 dB(A) [16].

10.5.2 Ray Tracing Modeling

It is not uncommon for production halls to feature shed-like roofs so as to benefit from natural lighting. More to the point, there often is much diffusion to be found on the walls, and even inside the hall, due to numerous fixtures (e.g., shelves, production equipment) being present.

A ray tracing model can help tackle those issues. Of course, compared to a Sabine model, things are slightly more complicated, as the dimensions and location of each surface must be entered [17, 18].

Regarding diffusion, an original solution has been proposed by INRS [17]: The sound ray is simply reflected in a random direction to take into account diffusion. Experience shows that it does work [19]. More to the point, some of those models have now started to include diffraction effects [20].

10.6 MEASUREMENTS

Looking at the aspects listed Section 10.1, there are a few points that may need verification through measurements:

- The sound power level of pieces of equipment (or the sound pressure level at the driving station)
- The sound propagation characteristics of the premises
- The sound exposure level of the workers

Sound power level measurements have been covered in several standards (e.g., the ISO 3740 series [21–23]). One must stress that the sound power level is heavily dependent on the conditions of operation of the equipment; it is necessary to follow the standard or the relevant contractual guidelines regarding this assessment.

Sound exposure level measurements can turn out to be real detective work if carried out properly. Basically, one must find out, per worker (or at least per group of worker), what his or her noise exposure level is. This is not a task to be taken lightly, as most of the legal obligations rely on this assessment. Standard EN/ISO 9612 gives a complete methodology of measurement as well as a list of things to be observed and reported on [2]. Please kindly note that while a good part of the monitoring job can be performed by the local person in charge of safety and sanitary matters, the actual definition of the measurement plan should be carried out by really qualified personnel [25].

Propagation measurements can be carried out according to ISO 14257. It must be stressed that the sound source used in such tests is omnidirectionnal. More to the point, one should keep in mind that a minimal distance (i.e., 3 m) should be kept between the omnidirectionnal sound source (or the measurement path for that matter) and any obstacle (e.g., wall).

10.7 EXAMPLES

10.7.1 Noise Reduction

In an untreated production hall there was a large machine generating a sound level of 85 dB(A) at the workstation. In addition, there also were 20 machines individually generating a sound level of 72 dB(A) at the same location.

When planning an improvement of work conditions, the management was quick to point out that a serious noise reduction effort was needed on the large machine. An expensive enclosure was custom built by a contractor with a planned noise reduction of 20 dB(A). On commissioning, it was discovered that the sound level at the workstation was still 85 dB(A) due to the noise contribution from the smaller machines.

Lesson Learned: Do take all the noise contributions into account, as numerous low-level noise sources may add up to represent a significant contributor.

10.7.2 It Would Have Been So Simple

When a highly sophisticated maintenance center was designed, the end user called for a suitably bright architect and his usual structural engineer to draft the relevant drawings. While everybody's attention was focused on a huge facility of this center, nobody paid any attention to smaller spaces that would eventually be easy to care for should the need arise.

It turned out that absorptive treatment really was needed in those small spaces; unfortunately, a mobile crane had been installed under the upper slab and there was not enough clearance left to apply the usual cheap absorptive baffles. This eventually led to some expensive corrective work involving the fixation of absorptive materials under the upper slab.

Lesson Learned: Noise control measures are costly when done afterwards!

10.7.3 Phantom at Work

A labor ministry inspector had pointed out a few potential problems in a brand new factory, and a noise control engineer was hired to investigate. Observing a significantly high noise level in an apparently deserted workshop, the acoustician was told, "Don't bother—there never is anybody in there as everything is automated." He then innocently asked why in such a dusty, supposedly unoccupied space there were two rather clean seats. Dumbfounded, the foreman admitted that while normal operation supposed that no human presence was required, numerous malfunctions in the process often prompted the lengthy presence of two workers over long time spans.

Lesson Learned: Observe and do not take for granted what was prescribed or announced.

10.7.4 Short-Duration Noise

Due to developing hearing impairment, an old worker had been posted to a rather quiet workplace. When a consulting engineer was tasked with the assessment of noise level exposure of workers, he had this particular worker fitted out with a dosimeter. In addition, he routinely performed sound level measurements in that workshop without ever breaking the 70 dB(A) mark. However, at the end of the day, the dosimeter gave an $L_{EX,d}$ value well over 85 dB(A). When asked about his activities, the worker merely answered he was doing over and over the same tasks; that was confirmed by his colleagues. Looking further into the matter, the acoustician eventually discovered that the 120 dB(A) mark was reached for 3 min at the beginning of each shift.

Lesson Learned: When in doubt, always observe and do not blindly trust the answers.

10.7.5 Short-Duration Task and Easy Noise Control

A company was willing to improve work conditions in its workshop on the grounds that production quality would then improve. The first year, it improved the thermal insulation and introduced better HVAC. The next year, it looked forward to improve the acoustics. Due to the complicated production schemes and layout of the workshop, the noise control engineer had to explain that he could of course do all the diagnosis work on both the workstations and the equipment, but that would be costly, and he suggested teaching the rudiments of acoustics to the safety engineer, the tooling engineer, and a worker who was a member of the safety group. The idea was retained, and those professionals-turned-students were enthusiastic. One week later the worker came back having diagnosed that a piece of equipment forgotten by everybody was operated a couple of times per week on a short duration (i.e., a few minutes). It turned out that it was an old ultrasonic cleaner that could generate noise levels in excess of 115 dB(A) at the workstation, with disastrous results on the $L_{EX,D}$. Fortunately, due to the high-frequency contents of the spectrum, it was easy to custom build a small enclosure using high-density mineral wool panels such as those intended for ventilation duct construction.

Lesson Learned: Beware of noisy short-time events forgotten by everybody.

10.7.6 Beware of the Too Obvious

A big airline had a large maintenance facility built, including a large shed and small workshops. It hired a program drafter to specify the required surfaces and physical parameters, and it quickly focused on the large shed, which could house half a dozen planes for maintenance and repair. Actually, the airline was so nervous that while its construction department hired an acoustician, the operations division hired another one, with both working on the shed. Both professionals remarked that the drawing excerpts of the hangar showed some much smaller spaces close to it labeled "metal workshop" or similar, but they were not listened to. On delivery, it turned out that nobody had cared for those small spaces with disastrous acoustic consequences. The spatial sound level decay in the metal workshop was 0.8 dB(A) per doubling of distance, and hammering on a metal plate generated noise levels close to the 100 dB(A) mark over several minutes.

Lesson Learned: Beware of program drafters and clients when they are not often in the field.

10.7.7 Enclosure for a Machine

Under pressure by the unions, the management of a production facility grudgingly decided to acquire an enclosure for a large production unit. Accordingly, they picked up a seemingly suitable enclosure in a manufacturer's catalog and ordered it without further haggling. On commissioning, the enclosure clearly featured the hoped for noise reduction over the 20 dB(A) mark. However, within 1 year the sound levels measured around the enclosure were back to their previous values, complaints regarding noise were rife, and management complained of having wasted money.

It was found that the enclosure had been delivered as a standard product. As no specific provisions had been taken regarding maintenance and access, both the production department and the maintenance department had to improvise. To avoid a complicated system of automatic access doors for the incoming or outgoing materials, production created a couple of apertures with a small absorptive tunnel connected to them; this certainly was better than nothing, but it nevertheless reduced the sound reduction of the enclosure. Next, maintenance discovered that access to a couple of sensitive areas was a bit complicated and had removable panels installed in front of those spots. Due to frequent interventions, some of those panels were often left dismounted for days and the sound reduction of the enclosure was further reduced. Last, production decided that the absorptive tunnels were prone to create stoppage and removed them. In the end, the sound reduction of the enclosure was a mere 5 dB(A).

Lesson Learned: Designing an enclosure requires time and numerous inputs from both production and maintenance; more to the point, the workers assigned to the particular process to be outfitted often have a couple of tricks up their sleeves that may not figure in the company's manual of practice. Ignoring their input will lead to failure.

10.7.8 Such a Simple Enclosure

An old yet useful piece of equipment was generating high levels of low-frequency sound in a workshop. Actually, the sound levels at the nearest workstation were found to be close to 80 dB(A). It was eventually decided that the simplest way to deal with it was to provide an enclosure. Accordingly, a seemingly suitable enclosure featuring a 10 dB(A) reduction was ordered from a manufacturer's catalog. However, on commissioning, the noise levels at the nearest workstation had not been reduced by more than 2 dB(A).

When the manufacturer was called for questioning, he quietly pointed out that nobody had presented him with the sound power level spectrum of the offending equipment; the advertised 10 dB(A) reduction was true of a pink noise, but certainly did not apply to low-frequency sound!

Having eventually discarded that industrial enclosure, the management opted for a heavy custom-built enclosure made of plain concrete blocks and with an acoustic door. This was satisfactorily implemented as far as acoustics were concerned. The maintenance department quickly found that the equipment was overheating inside the enclosure and considerably added an extraction fan; the corresponding aperture ruined the noise reduction, and the roar of the fan added some more noise in the vicinity.

Lesson Learned: When specifying an enclosure, call for the help of an acoustic specialist, or at least for somebody truly knowledgeable in these matters. Do not forget to exchange with the end users too, as they probably know from day-to-day operation what is practically feasible.

10.7.9 Cabin for Personnel

A company looking forward to implementing better work conditions decided to have control cabins in some noisy areas. More to the point, those large cabins were also to feature a few private spaces, allowing off-duty personnel to rest. A specialized manufacturer was duly contacted and asked to produce the required cabins as based on a standard design without any design team drafting a list of requirements on the users' side.

When the cabins were delivered, the users were flabbergasted to discover that the sound level inside the premises was close to 80 dB(A) due to their ventilation system. Asked about it, the manufacturer simply remarked that this was a normal noise level for an industrial cabin, and at any rate, nobody had requested him to aim for a lower level. When inevitably the case went to justice, the court noted that no requirement list had indeed been forwarded to the manufacturer, who had simply followed the customs of such a construction, and the users were left at stage 1.

Lesson Learned: Always prepare a requirements list, and when in doubt, call a specialist to assist in this task.

10.7.10 Workshop for Light Tasks

For the maintenance of electronic and computer equipment, a company rented a space in a building supposedly designed for multipurpose activities, with a standard 2.55 m height under the ceiling. In order to have maximum flexibility in the fitting of the space, the developer had opted for a plasterboard ceiling that permitted the implementation of a removable partition anywhere wished for by the client.

Within a week the workers complained about the space being too noisy; their anger grew when management performed a reverberation time measurement and declared the space fit for the working conditions. The acoustician who was hired next found that indeed the reverberation time was satisfactory, being under the 1.0 s mark, but the spatial sound level decay was a mere 0.8 dB(A) per doubling of distance.

The problem was eventually solved using absorptive material glued under the ceiling.

Lesson Learned: Reverberation time compliance is not sufficient to guarantee good acoustic conditions in the workplace. Incidentally, please note that gluing an absorptive material under the ceiling may require discussions with the firemen.

10.7.11 Assessing the Spatial Sound Level Decay

In France, the assessment of the spatial sound level decay is compulsory [12] in new or refitted workshops where the noise exposure level may reach or exceed 85 dB(A). As no standard was yet available for the task, a technical instruction was appended to the legal text. It basically stated what is to be found in ISO 14257 [10] applied to an omnidirectional sound source on the floor with measurement points 1.2 m high taken at distances that were multiples of 3 and 4 m. Quickly enough, there were disagreements between people assessing or measuring the spatial sound level decay. It was found (the hard way) that the sound source for such testing really had to be omnidirectional, and that the location of measurement or computation points was important.

10.7.12 Assessing the Noise Exposure Level of Workers

At the end of the 1970s a large industrial group in France decided to assess the noise exposure situation of its workers with regards to the legal target requirements (i.e., 85 dB(A) over an 8 h shift in those days). The medical and safety people of the various plants involved were tasked with this assessment. When the results eventually came, the management was dumbfounded: a plant reported 15% of its workers overexposed, while another, featuring the same equipment and layout, reported 85% of its workers overexposed! When the resulting row eventually cooled down, it turned out that the first plant had not taken into account some noises deemed accidental, while the other systematically did. This eventually led to the draft of a French standard [24] to prevent the occurrence of such disparities by making sure the same methodology was followed.

REFERENCES

1. Directive Européenne 86/188/CEE du Conseil du 12 mai 1986 concernant la protection des travailleurs contre les risques dus à l'exposition au bruit pendant le travail, *Journal Officiel de la République Française*, 137, 28–34 (1986).
2. EN/ISO 9612: *Acoustics—Determination of occupational noise exposure—Engineering method*, Geneva, 2009.
3. Décret n°2006-892 du 19 juillet 2006 relatif aux prescriptions de sécurité et de santé applicables en cas d'exposition des travailleurs aux risques dus au bruit et modifiant le code du travail—Deuxième partie : Décrets en Conseil d'Etat) (Decree on safety and health prescriptions applicable for exposure of workers to noise, in French), *Journal Officiel de la République Française*, 166 (2006).
4. HSE 2205: *Control of noise at work regulations*, http://www.hse.gov.uk/noise/regulations.htm.
5. Occupational Safety and Health standard 1910.95, U.S. Department of Labor, Washington, DC, 2013, https://www.osha.gov/law-regs.html.
6. Amendments to noise requirements in the regulations for industrial establishments and oil and gas—Offshore, Ministry of Labour of Ontario, Ottawa, 2007.

7. Canada Occupational Safety and Health Regulations, SOR/86-304, Section 7.4(1)(b), Canada Labour Ministry, Ottawa, 2006.

8. EN/ISO 11690-1: *Acoustics—Recommended practice for the design of low-noise workplaces containing machinery—Part 1: Noise control strategies*, 1996.

9. C. Hansen, *Noise control—From concept to application*, Taylor & Francis, London, 2005.

10. EN/ISO 14257: *Acoustics—Measurement and parametric description of spatial sound distribution curves in workrooms for evaluation of their acoustical performance*, Geneva, 2001.

11. EN/ISO 11690-3: *Acoustics—Recommended practice for the design of low-noise workplaces containing machinery—Part 3: Sound propagation and noise prediction in workrooms*, 1999.

12. Arrêté du 30 août 1990 pris pour l'application de l'article R. 235-11 du code du travail et relatif à la correction acoustique des locaux de travail (Arrest pertaining to the acoustic treatment of work spaces, in French), *Journal Officiel de la République Française*, 244 (1990).

13. ISO 17624: *Acoustics—Guidelines for noise control in offices and workrooms by means of acoustical screens*, 2004.

14. EN/ISO 11690-3: *Acoustics—Recommended practice for the design of low-noise workplaces containing machinery—Part 3: Sound propagation and noise prediction in workrooms*, 1999.

15. K. Pawlas, J. Grzesik, Efficiency of ear protectors in laboratory and real life tests, *International Archives of Occupational and Environmental Health*, 62, 323–327 (1990).

16. P. Danière, D. Robinet, *Sources sonores de référence—3ème partie: Décroissances sonores en atelier—Corrections de la DL (Reference sound sources—3rd part spatial sound level decays in workshops—Corrections of DL)*, MAV DT-379/PD, INRS, Nancy, 1998.

17. A.M. Ondet, J.L. Barbry, *Prévision des niveaux sonores dans les locaux encombrés (Prediction of sound levels in fitted rooms)*, NST0052, INRS, Nancy, 1984.

18. F. Probst, Prediction of sound pressure levels at workplace, presented at Acoustics 2012 Proceedings, Nantes, 2012.

19. A.M. Ondet, J.L. Barbry, *Acoustique prévisionnelle-Modélisation de la propagation dans les locaux industriels encombrés à partir de la technique des rayons-logiciel RAYSCAT (Predictive acoustics—Modeling of sound propagation in fitted industrial rooms using ray tracing—RAYSCAT program)*, NST0067, INRS, Nancy, 1987.

20. P. Chevret, J. Chatillon, Sound level prediction in open spaces: Implementation of diffraction in Rayplus software, presented at Acoustics 2012 Proceedings, Nantes, 2012.

21. ISO 3740: *Acoustics—Determination of sound power levels of noise sources—Guidelines for the use of basic standards*, Geneva, 2000.

22. ISO 374: *Acoustics—Determination of sound power levels and sound energy levels of noise sources using sound pressure—Survey method using an enveloping measurement surface over a reflecting plane*, Geneva, 2010.

23. ISO 3743-1: *Acoustics—Determination of sound power levels and sound energy levels of noise sources using sound pressure—Engineering methods for small movable sources in reverberant fields—Part 1: Comparison method for a hard-walled test room*, Geneva, 2010.

24. AFNOR S31084: *Acoustique—Méthode de mesurage des niveaux d'exposition au bruit en milieu de travail (Methods of occupational noise exposure measurement*, in French), Paris, 1984.

25. L. Thiery, P. Canetto, *Evaluer et mesurer l'exposition professionnelle au bruit (Measuring and assessing the professional noise exposure*, in French), Document 6035, INRS, Nancy, 2009.

Chapter 11

Educational Facilities' Performance and Lecture Halls

11.1 INTRODUCTION

Under such a name, quite a number of rooms and halls are to be found in educational facilities. As implied by such a name, while they usually are designed for a specific purpose (e.g., teaching, exercising, small performances by nonprofessional performers) and adapted to some extra uses, they often are to be found in small educational facilities or townships.

11.2 REQUIREMENTS

11.2.1 Foreword

As with any project, there usually exists a program: What does the end user (and the payer too!) actually want? This means the various requirements must be identified and the relevant acoustic objectives stated. In today's spirit of sustainable development, one must be conscious that a team effort (i.e., architects, structural engineers, HVAC, and acoustics, just to name a few) occurs and make sure that the various solutions that are considered at the design stage are compatible with each other. Acoustics often is part of a global problem that can only be solved by a complete design team fed the relevant data by the end user.

There are standards and regulations pertaining to the acoustics of classrooms and teaching spaces [1]. An analysis and comparison of the relevant regulations is given in a paper by Evans [2].

11.2.2 Classrooms

The requirement is pretty clear: The students must be able to hear clearly what the teacher says. Also, the teacher must be able to understand what the students are saying.

Yet some effort regarding noise control may have to be exercised according to the type of class; for example, small children can be quite agitated, so extra-absorptive treatment must be considered, especially with exercise classes.

There is a balance to achieve between the minimal acoustic absorption required, so as to guarantee reverberation control, and the necessary minimal spatial sound level decay needed, to ensure that the teacher's voice does reach the back rows of students. This is a reason for several regulations and standards that impose both a lower and an upper limit on the reverberation time (RT) value (e.g., [3]).

It is also necessary that the students at the back rows are not unduly annoyed by the noise coming from the classroom next door. This is not that easy an exercise, as there often are doors communicating from one classroom to another for both safety and operational reasons. A 40 dB target really is a minimum, with some regulations allowing a slight reduction of the required sound insulation between classes when there is a direct communication door (e.g., [7]).

11.2.3 Meeting Rooms and Offices

The requirement regarding a meeting room is quite similar to that of a classroom: Participants around the table must be able to hear distinctly what the speaker is saying or what the soundtrack from the audio/video system is playing. Also, participants must be able to understand each other when a debate is held. This means that like in a classroom, there is a balance to achieve between the minimal acoustic absorption required, so as to guarantee reverberation control, and the necessary minimal spatial sound level decay needed, to ensure that the speaker's voice does reach the farthest participants. Echo prevention has to be performed, and this often prompts for two adjacent absorptive walls, so as to prevent any risk of flutter echo.

Yet some effort regarding noise control may have to be exercised. To start with, there is the noise from the mechanical services of the building. But there also will be the noise from the ventilation of the audio/video system. The latter means that procurement of the equipment must be carried out with a clear acoustic target.

Sound insulation must be designed according to the needs. While a 38 dB sound insulation is a bare minimum between a meeting room and spaces of the same service, values of at least 45 dB are required for privacy purposes, and 55 dB for security purposes.

Some regulations require the sound insulation between offices to be the same as between classrooms (e.g., [7]). While this is a convenient basis for preliminary design, one must not forget that some offices (e.g., director's and doctor's) may need privacy with regards to other spaces, which means a target of 45 dB at the very least.

11.2.4 Lecture Halls

The requirements are the same as above. In addition, there usually is an additional requirement regarding the visibility of the board or screen.

Due to their higher status, lecture halls are usually treated to better standards, with a sound insulation with regards to other spaces that is at least 5 dB over the one required for regular classrooms. Due to their larger capacity, they may feature a sound reinforcement system (whose correct operation will usually require an absorptive treatment of the hall), or at least a suitable acoustic treatment, such as absorptive rear and side walls, and partly absorptive ceiling, with part of it designed to act as an acoustic reflector to help propagate the voice of the speaker to the audience. More to the point, more and more often they feature audio and video systems that may rival those installed in the local performance hall. Care must be applied to dialogue with the end user (what does he actually want?) as well as with the architect and the stage engineer (yes, there probably will be one assigned to the job, if only to define visibility curves from the seating as well as the audio/video system). Such an exchange may help prevent serious problems later, such as improper location with regards to other spaces.

11.2.5 Performance Halls

The performance halls considered here are small facilities within educational facilities with little or no stage equipment. For larger venues, refer to Chapters 12 to 15.

To start with, there is a visibility requirement, as the stage or the screen must be properly observed by the audience. This often induces a slightly stepped audience.

Next, there is a speech intelligibility requirement. In order to achieve a limited freedom of RT range, it is feasible to design a hard wall room in which the ceiling over the stage area will be reflective and slightly inclined, so as to send acoustic energy to the middle of the room, and the back wall will be absorptive, so as to prevent the occurrence of an echo. The remainder of the ceiling will be absorptive in the last third of the room and reflective in the middle. (Note: Should the hall be used with a flat floor without seating, then this ceiling must be inclined by at least 5°; otherwise, there will be a flutter echo.)

11.2.6 Restaurants

Restaurant and catering acoustics is one of the subjects of Chapter 7. Here are just a few basic reminders: To start with, restaurant and catering areas are supposed to be places where one is supposed to enjoy some relaxation at midday break, and this implies that there is not too much noise. This will have consequences on the kind of materials used, but also on the layout of the place (i.e., one cannot afford a dish cleaning room opened to the dining area; more to the point, it is advisable to keep the heavy kitchen activities, implying the noise of refrigerating units and exhaust systems, far enough away from the dining area too). It will even have consequences on the operations: Past a reasonable number of seats in the dining room, everyone will be in the direct sound field of another talker, and it will not be possible to carry out noise control measures [4]. This latter point will probably be a major source of discussion between the acoustician and the end user.

11.3 ACOUSTIC TARGETS

To start with, any project starts with a program. This is especially true of the classrooms and performance hall: What does the end user (and the payer too!) actually want? This means the various requirements must be identified and the relevant acoustic objectives stated. In today's spirit of sustainable development, one must be conscious that a team effort (i.e., architects, structural engineers, HVAC, and acoustics, just to name a few) occurs and make sure that the various solutions that are considered at the design stage are compatible with each other. Acoustics is often part of a global problem that can only be solved by a complete design team fed the relevant data by the end user.

There are some legal requirements regarding the sound insulation between classrooms (typically in the range of 45 to 50 dB) and the reverberation time (typically in the range of 0.4 to 0.8 s) or the minimal sound absorption area in the passageways. In addition, there usually are legal requirements regarding the noise generated by mechanical equipment (typically in the range of 30 to 35 dB(A) for permanently operating equipment). Those requirements apply to classrooms, but also to such spaces as offices, restaurants, workshops, infirmary, and the like. They may also complete some other regulations in the specific context of educational facilities. For example, in the French regulations, a D_{nTA} sound insulation value of 50 dB is required between a workshop and main spaces such as a classroom or an office [7], while the workshop must also comply with the occupational noise control requirements [8].

Do not forget that in addition to the legal requirements, the end user may have some more stringent requirements that are part of the contract.

11.4 A FEW BASIC RULES

First, one has to think of the sound insulation. The façade must be designed so as to prevent intrusion from outside noise. This may have some implications regarding the ventilation of the premises (can it be performed by merely opening the window, or is something more complicated in order?).

Next, partitions and doors must be designed in order to comply with the acoustic objectives. One must not forget that other constraints (e.g., fire safety and structural integrity) will apply too. One of the questions will probably be to decide whether direct communication doors are really needed between classrooms (they often are wished for by the staff, as they enable a single staff member to oversee several classrooms, and sometimes they may also be necessary as emergency ways out, but they are always a weak point in the sound insulation between classrooms).

Absorptive treatment must be applied. As pointed out, the basic rule is usually to try to end up with a RT between 0.4 and 0.8 s in the classrooms according to the regulations in force. Why that? Actually, one wants good intelligibility in the classroom (hence the upper RT limit), but one also wants the students farther from the teacher to be able to hear his voice at an understandable level (hence the lower RT limit that will ensure that the spatial sound level decay is not too big). This means that a ceiling with only a 0.70 absorption coefficient value will usually do the trick. However, when it comes to the restaurant and catering, it usually turns out that the more absorption in the space, the better it is for the acoustics. This may mean that should the end user want to use the dining room for other purposes (e.g., as an additional multipurpose room), some extra fittings may turn out to be necessary, but do not remove absorption from the space for the sake of an additional activity! In addition to providing absorption, one must clearly state that it is acoustically not acceptable to stuff as many students as possible within the dining room, as too noisy a situation will occur. This probably means that several dining services must be scheduled by the end user. The situation in the gymnasium is interesting enough: One wants to be able to express oneself, yet the teachers want to avoid a sore throat and drumming ears at the end of their shift! This means that a serious absorptive treatment is needed, using shock-resistant materials, on the ceiling as well as on two adjacent walls.

The noise of mechanical services must be controlled. More to the point, the HVAC network may be a path of sound transmission. This usually means that the supply and return ducts must be located in the corridor's ceiling, with silencers to the branch serving a room. It must be stressed that such a scheme must be announced well in advance, as there may be some implications regarding the amount of space needed as well as the fire safety precautions involved.

11.5 TYPICAL SPACES

11.5.1 Kindergarten School

In kindergarten schools, emphasis is usually put on safety. This means that precautions are implemented on the floor covering (usually a resilient floor covering in order to reduce the risk of a pupil getting injured when falling) and on the doors (usually wide resilient covering seals are used to prevent finger wounds). The latter means that the sound insulation may be reduced compared to other kinds of schools.

Sound absorption is usually provided on the ceiling and on part of the walls, as the emphasis is on noise reduction, with the teacher moving close to the groups of pupils according to the need.

11.5.2 Primary School

Primary schools usually follow the design rules of regular schools. This means that the sound insulation between classrooms typically is in the range from 40 to 48 dB, which usually calls for 50 to 55 dB partitions and 40 dB doors. A smaller sound insulation target is usually sought between classroom and corridor (e.g., 30 dB). The acoustic treatment of standard classrooms usually calls for a mildly absorptive ceiling in order to comply with a reverberation time objective in the range of 0.4 to 0.8 s.

11.5.3 College

The acoustic rules for classrooms are similar to the above. In addition, there are usually a few specialized classrooms and exercise rooms that require higher acoustic performance, such as workshops and music rooms, where a sound insulation of 50 to 55 dB is usually required with regards to other spaces. The acoustic treatment of such spaces is usually quite specific (e.g., the workshops feature an absorptive treatment complying with the targets of minimal spatial sound level decrease on the premises as defined by occupational noise control regulations (e.g., [9]), and the music rooms feature an acoustic treatment as per the recommendations of the culture authorities). In addition, some minimal dimensions may be required by those recommendations (e.g., a 4.5 m free height in the dance room).

11.5.4 University Theatre

University theatres can prove a bit complicated, as they may serve a large range of functions: Apart from being used for regular courses, they are often called on as conference facilities for scientific congresses, or as secondary performance halls for recreational and educational purposes. This means that they must be reasonably well insulated to enable simultaneous use of the various spaces of the building. Apart from an envelope enabling a 55 dB sound insulation to be reached, they usually are accessed through air locks (apart from better sound insulation than a single door, this prevents the undue intrusion of light and noise during courses or performances).

There are a few auxiliary spaces associated with such a theatre: To start with, there will be a projection booth that usually doubles as a technical booth for the control of lighting and sound systems that is located at the back of the theatre (though most controls can be operated from the speaker's position). There also is a waiting area for the speaker and his eventual assistants.

The typical acoustic treatment calls for an absorptive back wall (in order to prevent the occurrence of echoes) and mildly absorptive or diffusive lateral walls (in order to prevent the occurrence of flutter echoes). The ceiling usually is fully absorptive on the last third of the theatre, partly absorptive in the second third, and reflective over the speaker's area. However, in order to prevent the occurrence of a flutter echo in this area, it is preferable to have a ceiling inclined by at least 5°.

Noise control of mechanical equipment in such a space is a significant issue if one wants to achieve reasonable intelligibility. This means that proper thought will have to be given to the design and implementation of the HVAC.

11.5.5 Restaurant

Restaurants are covered in Chapter 7. The one point that is emphasized here is that apart from the efficiency of the absorptive treatment of the dining hall, the noise levels in a school restaurant are highly dependent on the number of pupils per unit area, on the mood of the day, and on the menu (e.g., fries usually generating much more excitement than peas). In order to achieve a reasonably nice restaurant, the question of density of people must not be overlooked (but unfortunately, it is just too tempting to squeeze in those extra people) [4].

Some points have to be discussed with the end user: To start with, it is acoustically preferable to avoid a dish-washing room opened to the dining room, but quite often, part of the staff likes the contact with the students. It may be possible to have a large opening as long as there is a washable absorptive noise barrier in front of the opening. Also, refrigerated shelves can generate significant noise, and it is preferable to avoid their presence directly in the dining room. Last, an open kitchen can generate quite a bit of noise in the dining room due to the exhausts, so it is acoustically better to avoid such a scheme too. To sum up, there should be four separate spaces: an entrance, a distribution area, a dining room, and an exit, all with absorptive material on the ceiling.

11.5.6 Sleeping Quarters

Sleeping quarters are a feature of kindergarten schools and nurseries. A significant sound insulation value is usually required between exercise areas and sleeping quarters (at least 50 dB). This is typically achieved through the use of thick heavy separating walls (e.g., 20 cm thick reinforced concrete walls) with an air lock (please kindly note that as the doors are considered emergency exits, one must not need to exercise too much effort to open them).

However, it is sometimes admitted that when those sleeping quarters serve the adjoining exercise areas, only the sound insulation can be downgraded to a much smaller value (e.g., 35 dB).

The treatment of such quarters usually calls for an absorptive ceiling and a resilient floor covering.

11.5.7 Gymnasium

The gymnasium is usually located away from the other spaces in order to help reduce the eventual impact noise transmission and noise from activities to other parts of the building. Should that be impossible, a floating floor will probably be needed in the gymnasium. The sound insulation with regards to the outside is usually rather poor due to natural ventilation and natural lighting (especially taking into account that most of the time the constructive elements must be shock resistant).

Acoustic treatment of the premises is of primary importance. To start with, the teacher must not be overexposed to the noise from activities; more to the point, he must be able to give instructions without shouting his head off, while the students must not strain to understand him. This acoustic treatment must, of course, be shock resistant, and several manufacturers have such material in their catalog [5, 6]. A typical treatment will call for the ceiling (or the underface of the roof) and two consecutive walls so as to prevent any significant resonance.

Noise control must be applied to heating and ventilating (cf. Chapter 4) so as to limit the sound level value at 45 dB(A) at most.

Usually there are basic requirements (e.g., sound insulation with regards to such sensible spaces such as classrooms, e.g., 53 dB in France [7], and a maximum reverberation time value too, taken from the regulations and recommendations applicable to sport halls [10, 11]) that are based on the volume of the sport hall.

11.5.8 Workshops

Workshops of educational facilities are normally covered by the same regulations as professional workshops regarding noise control inside the premises (cf. Chapter 10). In addition, there usually are legal requirements regarding the sound insulation with regards to such spaces as classrooms (e.g., 55 dB in France [7]).

One must especially be aware of the need for noise reduction inside the workshop, but nevertheless it is necessary for the teacher to be properly heard by the students. In addition to the normal treatment, it is advisable to provide a small space where discussions can be had or phone calls taken without being hampered by the noise of the activities around.

11.6 EXAMPLES

11.6.1 Nursery

A nursery was fitted out inside an existing Parisian building. The acoustician first made a diagnosis to check whether the built environment was compatible with the acoustic objectives. Then he issued specifications regarding the internal partitioning of the nursery and its acoustic treatment. To his consternation, the end user rejected most of his requirements. It quickly turned out that the one (valid) concern of the end user was safety. Therefore, the idea of properly insulating the various rooms as recommended by the available legal texts was considered preposterous, as the staff had to be able to hear anything untoward happening.

The set of acoustical objectives and relevant prescriptions was modified accordingly.

Lesson Learned: Always check with the end user!

11.6.2 Kindergarten School

A new kindergarten school was built downtown. It featured 20 cm thick reinforced concrete floors and 15 cm concrete walls. The potentially noisy exercise rooms were located at one end of the building. The master idea behind the concrete construction was to benefit from the thermal inertia of the building. Therefore, acoustic absorption was provided in the rooms by means of absorptive elements under 60% of the upper floor. A full ceiling was applied in the corridors for both acoustic and architectural purposes. In addition, perforated or slotted wooden panels were used in some large spaces, such as the entrance halls and exercise rooms.

11.6.3 Primary School

A new primary school was built downtown. It featured 20 cm thick reinforced concrete floors and either 15 cm concrete walls or 12 cm plasterboard partitions mounted from floor to floor between regular classrooms. Reinforced concrete (22 cm thick) was used for the music room. An acoustic absorptive ceiling was applied in each classroom, as well as in the corridors.

The clinch came from the neighborhood: People were complaining about the noise from the mechanical equipment (especially the boilers), as well as from the shouts in the gymnasium and recreational areas. The mechanical equipment was duly fitted with silencers, and proper thought was given to the sound insulation of the gymnasium. This led to mechanical ventilation (so as to avoid a direct opening), and thick windows with glazing on the outside and a metacrylate on the inner side. The recreational areas were treated to a shock-resistant absorptive noise barrier and absorptive baffles on top.

11.6.4 College

A new college was built in a small town. It featured 20 cm thick reinforced concrete floors and either 15 cm concrete walls or 12 cm plasterboard partitions mounted from floor to floor between regular classrooms. Reinforced concrete (20 to 25 cm thick) was used for specific rooms, such as workshops and music rooms. An acoustic absorptive ceiling was applied in each classroom as well as in the corridors.

11.6.5 University

During a major urban renewal project, a university building was fitted out in a former industrial building. Floors were typically 20 cm thick reinforced concrete. Classrooms were partitioned using 14 cm plasterboard mounted from floor to floor, and an acoustic absorptive ceiling was applied in each classroom, as well as in the corridors.

The 400-seat lecture theatres were built using concrete walls and stepped floor, with the ventilation under the seats. They featured a rectangular shape with absorptive material on the back wall and the lateral walls, while the main part of the ceiling was reflective, horizontal over the audience, and slightly lowering to the front wall.

One lecture theatre was specifically earmarked for special events, such as congresses. While most of the features were similar to the other lecture halls, it featured a cylindrical internal shell made of perforated metal with a mineral wool behind. Most of the ceiling was absorptive (perforated metal again), as its operation was to mainly use the electroacoustic tools available.

All the rooms and theatres were treated to a well-balanced ventilation system, with the same amount of air being supplied or taken in order to avoid whistling noises around the doors.

REFERENCES

1. ASA, Classroom acoustics booklet, 2000, asa.aip.org/classroom/booklet.htm.
2. J. Evans, Acoustical standards for classroom design—Comparison of international standards and low frequency criteria, presented at Low Frequency 2004 Proceedings, Maastricht, 2004.
3. *Association HQE: Référentiel bureaux enseignement (High Environmental Quality Association: Office and teaching facilities reference booklet*, in French), Paris, 2009.
4. ADEME, Caractérisation et réhabilitation des *cantines (Characterization and rehabilitation of school restaurants, in French)*, Final report, presented at GIAC 1993, Paris.
5. Rockwool, Boxer ceiling, Documentation, 2012.
6. Eurocoustic, Acoustichoc ceiling, Documentation, 2012.
7. Arrêté du 25 avril 2003 relatif à la limitation du bruit dans les établissements d'enseignement *(Arrest pertaining to noise control in educational facilities, in French), Journal Officiel de la République Française*, May 2003.
8. Arrêté du 30 août 1990 pris pour l'application de l'article R. 235-11 du code du travail et relatif à la correction acoustique des locaux de travail (Arrest pertaining to the acoustic treatment of work spaces, in French), *Journal Officiel de la République Française*, 244 (1990).
9. Décret n°2006-892 du 19 juillet 2006 relatif aux prescriptions de sécurité et de santé applicables en cas d'exposition des travailleurs aux risques dus au bruit et modifiant le code du travail—Deuxième partie: Décrets en Conseil d'Etat) (Decree on safety and health prescriptions applicable for exposure of workers to noise, in French), *Journal Officiel de la République Française*, 166 (2006).
10. *Directives et recommandations (guide technique) pour l'aménagement d'installations sportives (Directives and recommendations (technical guide) for the fitting out of sport facilities*, in French), Service de l'Education Physique et du Sport, Canton de Vaud, Lausanne, 2012.
11. *Salles sportives—Acoustique (Sport halls—Acoustics*, in French), Afnor, St. Denis, 1992.

Chapter 12

Theatres and Cinemas

12.1 INTRODUCTION

Theatres and cinemas mainly revolve around a same concern: giving a chance to the audience to see the plot on stage or on the screen while hearing properly the corresponding speech and sound signal.

12.2 REQUIREMENTS

To start with, there is a visibility requirement, as the stage or the screen must be properly observed by the audience. This usually induces a slightly stepped audience.

Next, there is a speech intelligibility requirement.

As with any project, there usually exists a program: What does the end user (and the payer too!) actually want? This means the various requirements must be identified (e.g., cinema only, or cinema and theatre? How loud will the music be? etc.) and the relevant acoustic objectives stated. In today's spirit of sustainable development, one must be conscious that the design team (i.e., architects, structural engineers, stage engineer, HVAC engineers, and acoustician, just to name a few) works together and make sure that the various solutions that are considered at the design stage are compatible with each other. This supposes the end user has made clear his intentions first! Should this not be the case, it will be up to the design team to push the end user into either acquiring his own support team or discussing with the design team to elaborate a program. Remember, acoustics often is part of a global problem that can only be solved by a complete design team fed the relevant data by the end user. One may also be confronted with an existing historical place where the appearance of the place is an important factor. At any rate, the visual aspects will be significant and must not be overlooked for the sake of acoustics, while decent intelligibility must nevertheless be provided.

12.3 ACOUSTIC TARGETS

12.3.1 Cinemas

The cinema industry has standards that are applicable (e.g., [2, 3]).

In the olden days, the sound insulation requirements with regard to the outside were not that high, as there were no sound systems save for the ubiquitous piano. Gradually, sound systems were introduced and the frequency domain of interest was steadily enlarged. To give an idea of this evolution, in the 1970s the sound insulation of cinemas used to be evaluated with a 90 dB(C) pink noise emitted in a theatre, while the sound level in the neighboring projection halls should not exceed 35 dB(A) and NR30. Nowadays, this evaluation is performed with a 105 dB(C) pink noise emitted in a projection hall, while the sound level in the neighboring projection halls should not exceed 32 dB(A) and NC27. More to the point, some special effects are sometimes required. The film *Earthquake* introduced spectators and neighbors alike to new sensations in the low-frequency range.

Therefore, the fundamental questions will be: What would the end user like to do, and what will the project actually allow him to enjoy (e.g., regular cinema projection, surround, or even Omnimax projections)?

Now let's consider the internal acoustics of the cinema projection hall: With reference to the idea that one must be able to perceive the sound as coming from the place on the screen where the action occurs, this means that acoustic reflections must be avoided. This in turn means that absorptive treatment should be applied on the walls and ceiling.

The sound levels generated by mechanical equipment should not be a hindrance. This typically means a background noise level value in the 30 to 35 dB(A) range, together with a frequency limit contour (e.g., NC) according to the standards in force.

12.3.2 Theatres

In the olden days the sound insulation requirements with regard to the outside were not that high, as there were no sound systems save for the ubiquitous piano. Gradually, sound systems and sound effects were introduced and the frequency domain of interest was steadily enlarged. More to the point, some special effects are sometimes required. As early as 1935 the physicist R. Wood had been asked by a stage engineer to produce some low-frequency sound during a play, and he had used an extremely long pipe, which made the structure shake to the extent that nobody wanted to try it again!

In addition to the acoustic aspects, a theatre can be quite different according to the kind of performance and audience. The most basic theatres, with a typical seating capacity of 100, will feature a small stage with a couple of access doors on each side of the stage back wall, and a couple of lighting fixtures over the front of the stage. Those doors are used to access the lodges and also to use the corridor linking both sides of the stage. Quite often there is no real proscenium wall, only curtains, which means that the stage practically is within the same volume as the audience. Such basic facilities are often found in schools and small townships.

More developed theatres will usually have a larger seating capacity, and will feature lateral accesses to the stage, in addition to those in the stage back wall, as well as storing areas. They will also have a couple of gratings over the stage to help with the positioning of lighting equipment and the hanging of back scenes.

Proper theatre facilities will have a real stage tower used to hoist back scenes when needed. This means that the height of the stage tower is practically double that of the theatre. More to the point, the top of the stage tower usually features gratings where blocks and tackles can be installed where needed. In old facilities, the hoisting relies on human power, and in modern facilities, it is electrically operated, which means there usually is a motor room to be insulated somewhere. In addition, due to the sheer lighting power needed, there is an electric room with gradating devices to be found somewhere in a room to be insulated (which can be located either close to the top of the stage tower, where most gratings converge, or at the bottom of the stage).

Therefore, the fundamental questions will be: What does the end user want to do, and what will the project actually allow him?

Now let's consider the internal acoustics of theatres; one must perceive the sound as coming from the place on the stage where the action occurs. More to the point, the speech intelligibility must be carefully nursed (meaning acoustic reflections on the side walls should be avoided and acoustic reflections on the ceiling should not be too late). Yet, sound propagation from the stage to the farthest seats should be ensured. This means that a significant part of the ceiling should be reflective and oriented so as to ensure correct sound distribution over the audience to achieve good clarity and speech intelligibility. Eventually, one usually aims for a RT of 0.9 to 1.0 s in the theatre proper and 1.5 s in the stage tower [4]. In addition, the strength G should be in the 10 to 13 dB range, and the clarity C_{80} should be in the 5 to 8 dB range. As one is looking for good speech intelligibility, the ALc should be kept under 10% (and it normally will be if the RT ranges are complied with) [4].

The sound levels generated by mechanical equipment should not be a hindrance. This typically means a background noise level value in the 28 to 35 dB(A) range.

12.4 A FEW BASIC RULES

As with any project, a cinema or theatre project starts with a program: What does the end user (and the payer too!) actually want (e.g., which capacity, which activities, what interactions with the immediate environment?)? This means the various requirements must be identified and the relevant acoustic objectives stated. In today's spirit of sustainable development, one must be conscious that a team effort (i.e., architects, structural engineers, HVAC, and acoustics, just to name a few) occurs and make sure that the various solutions that are considered at the design stage are compatible with each other. Acoustics often is part of a global problem that can only be solved by a complete design team fed the relevant data by the end user.

12.4.1 Cinemas

12.4.1.1 Projection Booth

In the olden days the projection booth featured a regular cinema projector and all the mechanical equipment needed to store the assembled film roll and pass it through the projector. This meant that enough space had to be allocated for all this equipment. In addition to the clicking sounds of the film rolling, there was the purr from the exhaust fan cooling down the projector. In order to reduce the amount of light emitted by the projection booth into the projection theatre, there usually was only a small openable aperture with a thick glazing for the operator to check the correct situation in the theatre, and a small aperture with a high-quality glazing for the projector. Due to the fire constraints (implying that the glazing and frame assembly had been subjected to a fire test) and the optical constraints (implying that the image would suffer no distortion), one typically was left with an 8 to 10 mm thick special glazing.

By the way, the facility always had an operator whose task was to prepare and assemble the rolls of films, and if need be, repair quickly the film roll in case of breakage.

Good news: Nowadays, numeric projection no longer requires a real operator, and one does not need to worry about the noise of this individual or the mechanical noise from the projector. Now for the bad news: It has become fashionable to show to the audience the modern projection equipment, so a large projection window is now used. However, due to the significant noise from the projector (whose exhaust is much more powerful than before), better sound insulation is needed between the projection booth and the projection room. This usually leads to a double projection glazing, with the outer glazing inclined for both acoustic and optical purposes.

12.4.1.2 Projection Theatre

There are a few trends with the projection theatres. To start with, the back wall is always absorptively treated in order to avoid an echo generated by the front loudspeakers. Next, most of the lateral walls are absorptively treated too. The front wall (which is largely behind the projection screen) is also absorptively treated.

In the olden days, the ceiling used to be reflective (e.g., plaster). Nowadays, there clearly is a need for an absorptive ceiling, at least over the flat area in front of the screen, as it is more and more fashionable to provide the audience with a small introductory conference. More to the point, it is not unusual for some operators to lease a cinema theatre for conference purposes. Unless an inclined reflective ceiling or an absorptive ceiling is used, there will be a flutter echo in this area. The other parts of the ceiling will be treated according to the needs, but one may care to note that in a normal-sized projection theatre, there is a strong need for low-frequency absorption, so a plasterboard ceiling may nicely provide that without adding extra medium- and high-frequency absorption.

Seats are of the normal performance hall type. The floor covering is usually carpet. Some operators favor plastic floor covering between the seats, but while it is easy to clean, it is also quite prone to generate whistling sounds when somebody moves a plastic-soled shoe on it, so acoustically it may be a problem.

The stepped flooring can be either precast concrete elements or thick wooden elements; the latter nicely adds low-frequency absorption in the theatre.

The walls of the cinema theatre can be made of thick (at least 30 cm) reinforced concrete walls, but they can also be made of thick (at least 45 cm) plasterboard partitions. In both cases, access and exit are performed through air locks, for both acoustic and lighting purposes.

Some projection theatres offer 3D vision. The basic idea is for the spectator to be able to match the visual action and the sound. This means that the inner envelope of the projection theatre must be highly absorptive so as to avoid any undue acoustic reflection. Two systems are currently used, both calling for a steeper stepped floor than conventional projection theatres: The first one requires the spectator to wear special glasses and the projection screen is partly cylindrical, while the latter does not require special glasses but calls for a hemispherical perforated screen behind which are located the air diffusers, the loudspeakers, and the absorptive treatment. In both cases, the thickness of the acoustic treatment is in the 50 to 100 cm range.

12.4.1.3 Public Spaces

While public spaces (e.g., corridors, entrance hall) are only temporarily occupied, they may deserve some acoustic treatment, especially as some of those areas may be designated as a waiting area for people with reduced mobility, and as such, some regulations [1] may require specific precautions. One will usually end up applying acoustic absorption on the ceiling so as to achieve an equivalent absorptive area of at least 25% of the floor area. Some extra absorption will be needed in such noisy areas as the image wall displaying the announcements for new films or the bar area.

Restaurants often are allocated to an independent operator who will have his own corporate image. It is advisable to introduce some acoustic absorption in those spaces too (e.g., applying acoustic absorption on the ceiling so as to achieve an equivalent absorptive area of at least 60% of the floor area).

When the cinema facility is located in a building housing other activities, especially dwellings, which is a situation not uncommon in Europe, a sore point will be the exits. They must be located as far as possible from the living spaces, bearing in mind that door slamming and eventual shouting are not the best of things at the end of an operating day when there are neighbors located close by! The trick usually is to apply acoustic absorption in the exit corridors and make sure that the doors do not slam violently.

12.4.1.4 Technical Areas

The technical area of a cinema can easily be sizable: There usually is one air handling unit (AHU) per room (as one does not want to lose the whole facility in case of a technical problem) plus one for the hall, one for the restaurant, and smaller units for the cold rooms of the restaurant and the sweet corner. While most of this equipment will stop after the operating hours, the cold room equipment will carry on. One must appropriately dimension the radiated sound power level of the equipment (including the relevant silencers and noise barriers).

12.4.1.5 Service Dwelling

Some facilities feature a small dwelling for the head operator or the director. While this individual will be awake during the operational hours of his facility, his family might not be. Therefore, it is highly necessary to provide some serious sound insulation (at least 65 dB) with regard to the activities in the facility.

12.4.1.6 Surround and Omnimax

The ultimate aim of the cinema as it is nowadays is to try to blend the optical image with the acoustical image. To achieve such a feast, there are a few techniques involved.

The oldest one had the projection performed on a hemicylindrical or even hemispherical screen. That screen was made of perforated metal, and the loudspeakers and HVAC terminals were hidden behind the screen [7]. Due to economics, the outer shell containing the facility usually was either a cylinder or a sphere, with corresponding risks of focusing that were solved through lots of absorptive materials (1 m thick is not uncommon for such purposes) located between the projection screen and the outer shell.

More recently some existing projection theatres have been outfitted with a surround system featuring loudspeakers on the lateral walls, back wall, and also ceiling [8]. In order to avoid creating a salient in the sight of the audience, recesses are needed in the ceiling, and this may seriously complicate matters with regard to sound insulation between projection theatres.

12.4.2 Theatres

12.4.2.1 Projection, Sound, and Lighting Booths

In the olden days, a theatre would feature a projection booth. Should there still be one in your project, cf. Section 12.4.1.1.

The lighting booth is used to control all the lighting fixtures of the theatre. There usually is some purring noise from the fans of the equipment. More to the point, as the booth might be operated open to the theatre, it is necessary to apply an absorptive treatment in this booth.

The sound booth is usually located next to it. Usually, the sound engineer will want to hear the natural sound of the theatre, so during operation it will often be opened to the theatre. An absorptive treatment will be needed in this booth to prevent the undue emission of noise from the cooling fans of the equipment (there definitely is some, though most of the amplifiers will usually be housed in a cooled soundproof enclosure, which may even be located farther away in the basement).

12.4.2.2 Theatre

There are quite a few trends with theatres. A theatre may be fan shaped or horseshoe shaped, and depending on the size, it may feature balconies. Most of the time the back wall is absorptively treated in order to prevent an echo; if it is not, then it will probably be diffusive. Next, the lateral walls are absorptively treated too (for good intelligibility, one avoids lateral reflections). The proscenium walls are often reflective.

A usual rule of thumb is to make sure that the volume does not exceed 4000 m³, with a maximum seating capacity between 800 and 1000, while limiting the volume per seat to 6 m³ [4].

The ceiling usually is reflective (e.g., plaster). Ideally, the shape of the ceiling is designed so as to help propagate the sound energy to the middle and the rear of the theatre. However, that shape often depends on the architectural wishes, as there will normally be a lighting bridge over the front of the stage that the architect may want to hide.

Seats are of the normal performance hall type. This means that the variation of acoustic absorption between the occupied or nonoccupied situation is no greater than 15%. The floor covering often is carpet.

In order to avoid a grazing incidence on the audience, which would be damageable for direct sound, an 8 cm clearance is required between rows. This usually is increased to 12 cm for visual purposes [6].

The walls of the theatre can be made of thick (at least 30 cm) reinforced concrete, but they can also be made of thick (at least 45 cm) plasterboard partitions. In both cases, access and exit are performed through air locks, for both acoustic and lighting purposes.

Balconies of a theatre should feature a height that is at least equal to its depth [4] or even greater than two times its depth [5].

Do beware of operators stating that they will never use amplified music! This is quite usual nowadays, and the sound insulation of the premises must be designed accordingly.

The stage tower usually houses quite a lot of machinery. It usually is absorptively treated (but nevertheless, remember that the reverberation time target is 1.5 s), bearing in mind that this treatment must be shock resistant to bear with the frequent handling of back scenes. Due to the amount of fire hazard material in this space, there usually are a few smoke exhaust trapdoors. They must be acoustically designed both to prevent the intrusion of external noise inside the stage tower and to prevent the radiation of noise from the show in the environment. A 45 dB trapdoor clearly is a minimum.

12.4.2.3 Public Spaces and Backstage Spaces

While public spaces (e.g., corridors, entrance hall) are only temporarily occupied, they may deserve some acoustic treatment, especially as some of those areas may be designated as a waiting area for people with reduced mobility, and as such, some regulations [1] may require specific precautions. One will usually end up applying acoustic absorption on the ceiling so as to achieve an equivalent absorptive area of at least 25% of the floor area. Some extra absorption will be needed in such noisy areas as the bar area.

When the theatre facility is structurally linked to other buildings, especially dwellings, which is a situation not uncommon in Europe, a sore point will be the exits. They must be located as far as possible from the living spaces, bearing in mind that door slamming and eventual shouting are not the best of things at the end of an operating day when there are neighbors located close by! The trick usually is to apply acoustic absorption in the exit corridors and make sure that the doors do not slam violently. An even worse point will be the unloading and loading of the back scenes before and after the performance; while the former usually happens in daytime, the latter often is performed after the performance, that is, in nighttime. There may be shouts, the beeping of the truck backing up to the loading area, the noise from the forklift. The loading/unloading area will have to be properly located with regard to the neighborhood; this often results in an enclosed facility.

12.4.2.4 Technical Areas

The technical area of a theatre can easily be sizable: There usually is one large AHU for the theatre, plus another one for the dressing rooms, one for the hall, one for the offices and cloakrooms, and smaller units for the sweet corner. While most of this equipment will stop after the operating hours, the cold room equipment will carry on. One must appropriately dimension the radiated sound power level of the equipment (including the relevant silencers and noise barriers).

12.4.2.5 Service Dwelling

Some facilities feature a small dwelling for the director. While this individual will be awake during the operational hours of his facility, his family might not be. More to the point, the living room often doubles up as a reception area for guests. Therefore, it is highly necessary to provide some serious sound insulation (at least 65 dB) with regard to the activities in the facility.

12.5 EXAMPLES

12.5.1 A Noisy Service Dwelling

A mid-sized cinema featured a service dwelling located close to the main projection booth (a rather common feature of older facilities in Europe). The sound insulation between those two spaces was not high performance, being only 48 dB, as it had been thought that the director would be awake during operational hours. When a new director was hired, tempers flared quickly with the head operator. The latter, being a few months shy of retirement, found a nasty vengeance by fully turning on the small loudspeakers in the main booth, and with deadpan face explaining to the director that this was normal procedure for mishearing people. The unfortunate director only found out the truth with the arrival of a new head operator.

Lesson Learned: You had better know the technique before facing an old hand.

12.5.2 A Small Township Cinema

A small township cinema seating 80 was built in a French municipal building also housing a gymnasium and a municipal garage. It featured one cinema theatre and its attendant projection booth. There were a small entrance hall and a bar corner. Due to the different operating hours, those activities did not turn out to be a hindrance for each other, and as the building was stand-alone, there was no neighboring problem.

In another town, a new 250-seat cinema and its attending projection booth and entrance hall were built. In order to prevent the transmission of noise to the neighboring dwellings, a sizable (10 cm) expansion joint was left around the building. The noise from the mechanical equipment was controlled using an enclosed plant room and silencers.

12.5.3 A Large Cinema Facility

A 5400-seat 22-projection cinema was built in Strasbourg, France, for the operator UGC. Due to the seismicity of the area, it was necessary to build the facility as structural blocks of two theatres side by side; in order to prevent an eventual toppling of the structure, the weight of the upper floor was kept down using a 20 cm thick reinforced concrete slab connected to a plasterboard suspended ceiling, and the partitions were made of 65 cm thick plasterboard partitions featuring a sound reduction index of 75 dB. All the mechanical equipment was located on the roof.

12.5.4 An Enlargement of a Large Cinema

A 1500-seat cinema facility was heavily refurbished in Bordeaux, France, by operator UGC. At the time of the renovation, the projection theatres had been designed with a larger capacity in mind; however, due to the fact that all exits were located on the same side, it was not deemed possible to reach such a capacity due to the safety regulations in force. However, 10 years later the township approached the operator: They wanted three small projection rooms devoted to art cinema, and they were ready to allocate a recently vacated building on the other side of the block. This meant that a new exit on another façade was available, and the capacity of the existing facility could be significantly increased.

While the existing façades of that building were kept, new floors were built; due to structural restrictions, they were only 15 cm thick, and a large plasterboard ceiling was implemented. The partitions were of the plasterboard on studs type. This enabled the successful construction of three 50-seat projection rooms.

12.5.5 A Small Township Theatre

A small township got itself a small theatre. It was a simple steel structure clad with metal panels in which there was a 200-seat theatre with a small raised stage complete with lateral spaces and a backstage area so as to allow some stage moves, though it did not feature any stage tower. The dressing rooms, mechanical equipment room, and back scene storage were all at the stage end of the theatre, while the entrance hall and adjoining space used as a bar area or a meeting space were at the opposite end. Absorptive treatment was applied on the ceiling of the auxiliary spaces and also on the walls of the back stage area, while the theatre was treated to absorptive walls and a reflective ceiling. Such a simple facility did serve the community well, though the light walls could of course not allow the use of powerful electroacoustics in the facility.

12.5.6 A Medium-Sized Theatre

A group of townships had reserved some downtown landscape for cultural activities. They first built a small multipurpose facility, but left vacant an adjacent space for future use. Several years later, this facility had been operating satisfactorily and had proven that there was a need for a more sophisticated facility. With some funding available, it was eventually decided that a 300-seat theatre could be built.

The new concrete structure was built with a 10 cm expansion joint separating it from the multipurpose existing building. It housed a 300-seat theatre complete with a stage tower.

The theatre was rectangular shaped (as this had the better fit in the available space), had absorptive treatment on the back wall and in the stage tower, while the ceiling over the audience was reflective using plasterboards, with a shape designed to enhance the propagation of natural voices from the stage to the audience. In order to provide some flexibility with the acoustics of the place, the side walls in the theatre were treated to mobile elements that could be either absorptive or diffusive; while this did not unduly affect the reverberation time (which would drop from 1.2 to 1.1 s in full absorption mode), it did provide lateral acoustic reflections that were enjoyed when a performance involved some music.

In addition, the building featured offices and a catering area for the staff.

Due to the proximity of dwellings, it was necessary to implement specific noise control provisions for the loading/unloading area. This resulted in an absorptively treated covered area where a full semitrailer truck could back in and be loaded or unloaded using a forklift after closing the garage-like doors.

12.5.7 A Large Theatre

A large theatre was built in a regional capital using subsidies from the ministry of culture and the regional authorities. The concrete structure featured a 400-seat theatre complete with stage tower and relevant stage equipment, as well as a smaller 100-seat facility with a slightly raised stage but no stage tower. The former was intended for real performances by professionals, while the latter was available for nonprofessionals and schools.

The larger theatre was slightly fan shapes (as this maximizes the sight lines), had absorptive treatment on all the walls and in the stage tower, while the ceiling over the audience was reflective using plasterboards, with a shape designed to enhance the propagation of natural voices from the stage to the audience.

The smaller theatre was treated to absorptive walls and fan shaped so as to minimize the distance between the stage and the audience. The side walls were reflective, but heavy stage curtains could be drawn in front of them to reduce the reverberation time and prevent the occurrence of side reflections. A reflective panel was suspended over the front of the stage in order to help propagate the sound from normal voice on the stage to the audience without hampering the lighting effects. Both theatres were served by the same entrance hall.

In addition, the building featured rehearsal rooms and offices, as well as a catering area for the staff and a restaurant for the public.

REFERENCES

1. Arrêté du 1er août 2006 fixant les dispositions prises pour l'application des articles R. 111-19 à R. 111-19-3 et R. 111-19-6 du code de la construction et de l'habitation relatives à l'accessibilité aux personnes handicapées des établissements recevant du public et des installations ouvertes au public lors de leur construction ou de leur création (Arrest dated August 1, 2006, defining prescriptions for accessibility by reduced mobility people to public receiving facilities, in French), *Journal Officiel de la République Française*, August 2006.
2. CST (Commission Supérieure Technique de l'Image et du Son) (French Technical Commission of Image and Sound), *Guide de l'exploitation cinématographique* (*Guide of opération*, in French), Paris, 2003.
3. THX, THX certified cinémas, http://www.thx.com/professional/cinema-certification/thx-certified-cinemas/.
4. Peutz, Standard guide for initial design, Internal document, 2004.
5. A. Gade, *Acoustics in halls for speech and music*, Springer Verlag, Berlin, 2007.
6. M. Long, *Architectural acoustics*, Elsevier Academic Press, Burlington, MA, 2006.
7. W.C. Shaw, J. Creighton Douglas, IMAX and OMNIMAX theatre design, *SMPTE Journal*, 97(3), 1983.
8. *Doby ATMOS specifications*, Issue 2, DOLBY Publication S14/27333/27842, 2014.

Chapter 13

Music and Concert Facilities

13.1 INTRODUCTION

Music and concert facilities of any type are usually meant to provide the audience and the musicians with a shelter from intrusions from the environment. They must enable the musicians to hear each other, while giving the audience a decent chance to look at the musicians and hear their part too.

The word *music* can cover quite a broad extend of sound pieces. Victor Hugo used to say, "Music is noise that thinks." There is a bit of difference in terms of noise levels and frequency range between the baroque instruments and the electroacoustical equipment of today, as well as in their preferred listening environments! Clearly enough, the corresponding facilities will be rather different, starting with their acoustic targets [1, 2].

13.2 REQUIREMENTS

13.2.1 Foreword

As with any project, there usually exists a program: What does the end user (and the payer too!) actually want? This means the various requirements must be identified and the relevant acoustic objectives stated. In today's spirit of sustainable development, one must be conscious that a team effort (i.e., architects, structural engineers, HVAC, and acoustics, just to name a few) occurs and make sure that the various solutions that are considered at the design stage are compatible with each other. While extremely important, especially in a concert hall, acoustics often are part of a global problem that can only be solved by a complete design team fed the relevant data by the end user.

There are a couple of rules that are often quoted by the end users (Warning: The end user often is the payer too, so watch your step!). To start with, any good hall has wooden paneling (this is obviously false, but you will have to explain it very diplomatically!). Next, a volume of 10 m³ per auditor must be provided. While this number can be used for predimensioning purposes during the programming phase, it is neither sufficient nor necessary to achieve good acoustics in a classical concert hall.

13.2.2 Concert Facility

There can be quite a broad variety of concert facilities, from the small 200-seat rather reverberant salon used for baroque music to the 5000-seat facility used for pop music. The former is usually reverberant (which is a help for music instruments that have difficulty producing significant sound levels, and provides spaciousness too!) and found in historical buildings or cultural centers, while the latter is a stand-alone construction rather heavily dampened in order to minimize acoustic reflections on the inner envelope of the facility (which is a help to the sound engineer of the concert).

To start with, there is a visibility requirement, as the stage must be properly observed by the audience. This usually induces slightly stepped audience seats. But there is a reverberation requirement, in terms of both reverberation time and spaciousness.

Unless there is a sizable budget for variable acoustics (in terms of both acquisition and operation personnel), the RT will inevitably be fixed. As the requirements are quite clear (a rather dead room for electroacoustic music, a quite reverberant room for baroque music, and an even more reverberant room for symphonic music), one will have to make a choice according to the wishes of the end user.

To start with, it is, of course, unthinkable to sit in a concert hall and hear the traffic noise outside due to weaknesses in the envelope of the hall. Does it sound ridiculous? Please have a look at Section 13.6.10.

In addition, there is a strong requirement on the sound insulation with regard to other spaces. A bit of advice here: One might be unhappy hearing the noise from the concert inside other rooms of the premises, but that is the problem of the end user—that is, as long as he willingly went for it. But it is out of the question for the neighbors to be submitted to the noise of music; to start with, this is considered to be an offense, as it comes under either the community control regulations or specific regulations pertaining to musical venues. Neglecting this aspect will sooner or later end in failure.

Beware of operators who say either they do not use electroacoustical instruments, or they introduce a sound-limiting device on the sound system. It is just too tempting to use the facility for such purposes for one evening, and then the legal fun will start. On pointing out that the sound insulation with regard to the environment is too weak for comfort, when being answered that there will be a sound-limiting device on the sound system, stand by to stop the project: Live musicians will not use such a system anyway!

13.2.3 Exercise and Rehearsal Rooms

Good news: The problem is much simpler when it comes to internal acoustics due to the fact that those are rather small rooms. Now for the bad news: Such rooms are seldom alone, and one has to design the sound insulation with regard to the neighborhood and also to the other rooms in the building. This usually results in numerous box-in-box constructions. If it is a brand new facility, it is acceptable; if it is a refurbished building, chances are high that it will not work for either acoustic reasons or safety problems.

Having taken care of the sound insulation, one now has to tackle the ventilation of the premises. If a rather low background noise is aimed for (e.g., because the operator wants the possibility to perform recordings), this means a low air speed, which translates into large ducts (yes, plural: one in and one out). At any rate, in order to satisfy the sound insulation requirements between rooms, there will be a need for sizable silencers too.

What should be the internal acoustics of the room? It is tempting to look for a rather deadened room. However, while this is of interest for professional musicians attempting to improve their playing skills, it can turn out to be quite nightmarish for the average beginner, who will be under the impression that he cannot develop the sound from his instrument. On the other hand, one must avoid any undue resonance in the room. This means that nonparallel surfaces will be in order.

13.3 ACOUSTIC TARGETS

13.3.1 Foreword

The acoustics of the place and the musical works performed inside are often quite linked one to each other. For example, in the first half of the 18th century German composer Johan Sebastian Bach was writing his organ pieces for rather small churches with lots of fittings; this resulted in a rather low reverberation time value that enabled him to practice a dense melodic writing. On the opposite end, his French contemporary Nicolas de Grigny was writing for large and bare churches; as the articulation in such spaces would be problematic, he resorted to a dense harmonic writing [3].

13.3.2 Sound Insulation

In the olden days, the sound insulation requirements with regard to the outside for community noise control purposes were not that high, as the sound power level of the music instruments of the time was not that high. There were no sound systems except the eventual symphonic orchestra. Gradually, sound systems were introduced and the frequency domain of interest was steadily enlarged. But even then, some facilities had been tackling the low-frequency range: An organ with a 32 ft tube can generate sounds down to 16 Hz, which are useful as a fundamental for creating sounds, and those were in existence in the 19th century.

Nowadays, modern sound systems can generate high noise levels over a very broad spectrum. This means that the sound insulation with regard to the outside must be treated accordingly. There usually are some legal requirements (e.g., in France, a decree on musical venues using amplified music [4]). In addition, care must be taken regarding the protection of the hearing of both the audience and the musicians.

Therefore, the fundamental questions will be: What does the end user want to do, and what will the project actually allow him to do?

13.3.3 Reverberation

The internal acoustics of a concert facility must be adjusted to its finality: When dealing with modern music generated through electroacoustical means, the basic idea is to avoid acoustic reflections on the walls and ceiling, which leads to an absorptive treatment of the concert facility. However, when dealing with classical music (implying natural acoustics), one looks for an enveloping sensation through numerous reflections and diffusion on the walls and ceiling (meaning acoustic reflections on the side walls must be provided, and acoustic reflections on the ceiling should be provided too). This has practical consequences: In order to benefit from the wall reflections, a shoe box volume is usually preferred for such purposes. A fan-shaped volume, while affording better vision of the stage to the audience, will not enable as much wall reflections. To give an idea of the reverberation time range, one will try to aim for 1.0 to 1.3 s for an amplified music facility, together with a clarity C_{80} greater than 10 dB. But for a symphonic hall, one will find 2.0 to 2.2 s for the occupied hall, together with a clarity C_{80} in the range −3 to +3 dB and a strength G in the 3 to 6 dB range. By the way, beware of generalities regarding baroque music: Even if it now is fashionable to execute such works in a small hall with a RT of circa 1.6 s, most of the baroque musical works were usually composed for a specific venue where conditions might differ a lot from those of another venue (and remember, in those days music would not be distributed as easily as now, which accounted a lot for differences between venues [12]. Let us say that 1.6 s may be a basis for the design, but the conductor (and the acoustician, if he is game for it) will probably end up doing some extra work looking for the right disposition to cope with the acoustics of his venue (e.g., using an absorptive surface behind the horns and exiling the brass at the end of the stage).

Sound propagation from the stage to the farthest seats should be ensured. This means that a significant part of the ceiling should be reflective and oriented so as to ensure correct sound distribution over the audience.

An important feature of concert halls is that the stage and the audience are in the same volume.

Looking at the internal acoustics of concert halls, one must perceive the sound as coming from the place on the stage where the instruments are; even then, one may distinguish between the pure classical works, where each set of instruments must clearly be distinguished, and the later works, where the orchestra is treated as a global mass.

13.3.4 Mechanical Noise

The sound levels generated by mechanical equipment should not be a hindrance. This typically means a background noise level value in the 25 to 35 dB(A) range for a regular concert hall, but can be higher for a pop music facility, where a 35 to 42 dB(A) noise level is often found acceptable. Actually, some specification books for such facilities do not even mention the possible problem of mechanical noise inside the facility [5].

13.4 A FEW BASIC RULES

13.4.1 General Rules

As with any project, a concert hall or a music facility project starts with a program: What does the end user (and the payer too!) actually want (e.g., capacity, shape, seated or standing audience, type of music to be preferably performed in the facility, interactions with the immediate environment of the project, etc.)? One might care to note that capacity is an important factor in terms of both acoustics (it may mean the implementation of balconies) and safety (due to the increased number of emergency exits), and it also means income. This means the various requirements must be identified and the relevant acoustic objectives stated. In today's spirit of sustainable development, one must be conscious that a team effort (i.e., architects, structural engineers, HVAC, and acoustics, just to name a few) occurs and make sure that the various solutions that are considered at the design stage are compatible with each other. Acoustics is part of a global problem that can only be solved by a complete design team fed the relevant data by the end user. Still in doubt about that? Have a look at Section 13.6.9 then!

The very first question is: What kind of music is the facility intended for? Modern music (e.g., pop, electroacoustic music) will need highly absorptive walls and ceiling, and will not crave reverberation. A classical music facility will need reverberation and lots of acoustic reflections, and this will reflect on the design.

Do beware of operators stating that they will never use amplified music! While they may be sincere in their saying this, experience shows that sooner or later somebody is going to introduce loudspeakers and amplifiers in the venue. Even if the venue is labeled "chamber music," sooner or later the day will come. Specific statements and prescriptions will have to be spelled out on this subject. At any rate, some classical instruments can be quite loud too (e.g., a French horn can generate up to 120 dB(A) during a fortissimo).

Do leave enough space for ventilation purposes. The smaller the space allocated to HVAC, the noisier the system will be. The air speed should not be greater than 2.5 m/s to avoid too much noise generation. A good strategy is often to try to have air distribution under the seats, as it is delivered at low speed where it is useful.

Variable acoustics can be sought when one wants to adjust the reverberation time according to the task. This can be achieved to a limited extent by using heavy velvet curtains and banners. There nowadays are some electronic systems capable of adding energy so as to reproduce extra acoustic reflections for reverberation time increase as well as better envelopment. While such systems usually are used for remedial purposes (i.e., the RT value was initially too low, or became too low following some modifications in the venue), they can also be used in new projects [6].

13.4.2 Amplified Music Facilities

With modern music generated through electroacoustical means, the basic idea is to avoid acoustic reflections on the walls and ceiling, as they would significantly complicate the life of the stage sound engineer. In addition, the typical audience to be found in such facilities is usually not the quiet type to be found in cozy opera houses. This will eventually result in an environment where the floor is bare concrete, the seats are either hard wood or thick plastic, and the walls are treated to thick layers of fibrous materials to achieve the required absorption in the low-frequency range. This is especially important in such facilities, as most of the time, in order to preserve the neighborhood, the envelope of the hall is highly insulating (e.g., a $D_{nTw} + C_{tr}$ value of 65 dB is quite common); more to the point, there are legal requirements regarding the sound insulation to be achieved (e.g., [4]), and those must be established prior to the building permit. Under such conditions, the low-frequency sound energy that usually goes through the envelope has to be dealt with through acoustic absorption. One should note that in order to achieve a reasonably flat reverberation time curve over the frequency range of interest, the use of materials featuring good absorptive characteristics in the low-frequency range and lesser characteristics in the higher-frequency range is recommended (as a reminder, the greater the volume, the higher the air absorption). This will typically result in the use of multilayered absorptive materials or fibrous materials protected by a perforated plate (by the way, do remember that due to the kind of audience, the material has to be resistant to various types of aggressions).

Access to the stage area will be made possible by a forklift, as most shows nowadays use such a convenience. This means that a large set of doors will be needed. More to the point, access to the facility will have to be provided through air locks, and the architect may well be left with the question of designing an extra envelope permitting the use of the bar and the toilets during the show!

Quite often in such facilities, the noise generated by the air handling system is not too much of a problem as long as it is designed with a 35 to 42 dB(A) target in mind. (Sounds like a lot? Well take into account the heat gain of all the audience plus the stage lighting and sound equipment, and you will see how much flow rate will be needed!) The one element that may cause problems is the smoke exhaust and fresh air supply to be used in case of fire. Depending on the local fire regulations, it will require silencers or trapdoors triggered by the fire detection system. In the absence of such equipment, you can bet that the noise generated inside the facility will find its way outside.

13.4.3 Classical Music Facilities

A concert hall for classical music (implying natural acoustics) needs to provide an enveloping sensation through numerous reflections and diffusion on the walls and ceiling. This means that acoustic reflections on the side walls must be provided, and acoustic reflections on the ceiling should be provided too. This has practical consequences: In order to benefit from the wall reflections, a shoe box volume is usually preferred for such purposes. A fan-shaped volume, while affording better vision of the stage to the audience, will not enable as much wall reflections. More recently, a vineyard terrace shape (introduced at Berlin's Philharmonie by architect Hans Scharoun and acoustician Lothar Cremer) subdivides the audience and provides more reflections from the dividing walls [7], while with most of the audience being seated close to the stage, the hall sounds comfortably loud.

Due to the need for hard reflective surfaces, concave wall and ceiling shapes had better be avoided. When this is not possible (e.g., because of an already existing facility), a diffusive treatment must be provided. A highly visible example of such a necessary treatment can be found in such an elliptical shape as the MC2 auditorium in Grenoble (cf. Section 13.6.2).

Sound propagation from the stage to the furthest seats should be ensured. This means that a significant part of the ceiling should be reflective and oriented, so as to ensure correct sound distribution over the audience. This also means that the use of balconies must be done with much care, or one will end up with dead spots. Typically, a height greater than the depth is sought [14],

In order to avoid a grazing incidence on the audience, which would be damageable for direct sound and also for reflections from the side walls, an 8 cm minimum clearance is required between rows. This is usually increased to 12 cm for visual purposes [15].

Eventually, for a symphonic hall, one will usually end up with a 10 to 12 m^3 volume per spectator, with a total volume ranging between 15,000 and 20,000 m^3. The same rules of thumb will limit the width of the hall to 25 m and its length to 50 m, while the height-to-width ratio will be kept under 0.7. By the way, should there be balconies (which can provide interesting diffusion and reflections), their proportion should be such that their height is similar to their length, as one will prefer the reverberated field over the direct field [13].

Noise from air handling systems can quickly be a problem in such facilities: In order to cope with the large number of spectators, one clearly has a preference to use many low-speed outlets located under the seats. Adding their sound contribution at a given seat, one can easily end up with a sound power level requirement under the NR15 mark. Under such conditions a laboratory test report of the outlet must be provided.

13.5 MODELING

Modeling is a nice tool, both for the specialist and for public information. Like all tools, it does have limitations and cannot be trusted blindly! While it is nice to establish a comparison between different situations—provided the model has been properly elaborated—it seldom comes up with an absolute value to be trusted and cannot be operated by nonspecialists.

13.5.1 Ray Tracing Modeling

Ray tracing using a computer does look the part. However, the physically minded can quickly understand that solving the wave equation inside a hall full of singularities such as balconies, coupled volumes, and funny-looking surfaces (including curved surfaces) is not going to be easy, not to say accurate. Nevertheless, ray tracing is rather quick to implement (especially if the architect has previously been talked into providing the relevant drawings suitably clean of unnecessary details) and can be used to pinpoint and correct such defects as echoes and inadequate reverberation time.

Some commercial programs now provide auralization modules that enable the user to hear how the hall under study will eventually sound.

13.5.2 Scale Modeling

Scale modeling may look old fashioned, but it always features a kind of magic to onlookers. Actually, one may discuss scattering objects for hours in front of a dubious client using some computer simulation results and images without much success, but having the client look at a scale model while being told the problem usually is quite effective, not to say awesome.

Scale modeling proves efficient when dealing with singularities such as balconies, coupled volumes, and funny-looking surfaces.

What scale should be used? Clearly enough, the smaller the scale, the higher the frequency range will be and the more complicated the measurements will turn out, as air absorption becomes quite a problem. Scales like 1/10 to 1/16 are rather nice and popular, as air absorption is not yet too much of a problem, but while the model is easy to work with, it also happens to be quite cumbersome. Experiments have been performed with cardboard 1/50 models [8].

13.6 EXAMPLES

13.6.1 IRCAM

The French Institute of Research and Coordination Acoustics and Music was defined as a part of a larger project for a contemporary art museum in the city of Paris by architects Mike Rogers and Renzo Piano. Among the various spaces of interest, it features a variable acoustics hall whose RT can vary from 1.1 to 4.4 s. This is obtained using a variable-height ceiling in three parts (so as to reduce the volume) that can be moved independently, while both the walls and the ceiling feature movable panels that will, according to the requirements, be either reflective, diffusive, or absorptive [9].

13.6.2 MC2

The Culture Centre of Grenoble (France) was inaugurated in 1968. Designed by architect André Wogenscky, it featured a music auditorium, a large theatre hall, as well as various other performance halls and their supporting rooms. A major rehabilitation project was performed at the beginning of the 21st century by architect Antoine Stinco.

Due to its elliptical shape, the 998-seat auditorium needed some highly diffusive treatment on the walls to prevent the apparition of echoes. The study of the treatment was performed first using a computer ray tracing model, and later on using a 1/16 scale model. The latter proved useful in many ways: To start with, it enabled the acoustician to perform an experimental study to optimize the shape of the diffusive surfaces while taking into account the coupled spaces of the galleries. Next, it also helped convince the architect and the client regarding the general appearance of the hall, while showing rather easily how the removal of part of the treatment would result in an echo.

13.6.3 Zenith of Dijon

The Zenith of Dijon (France) was designed by the French architect Philippe Chaix, who is considered a specialist in such facilities. It seats 5500 (with a full capacity of 7000 with standing spectators).

While the acoustic absorption was of course designed to cope with pop music, proper attention was given to the control of echoes and good intelligibility, as well as to the noise control of the mechanical equipment. Shortly after commissioning, an unhappy director called the acoustician to report that the particular visiting group was unhappy with the acoustics. Rushing on the scene, the acoustician was met by a grim director and the representative of the township, who was suitably anxious, as the payer of the project. To the consternation of both the acoustician and the town representative, the sound engineer in charge of the visiting group then explained that the acoustics were bad as any sound from the stage was fully heard and understood in the audience (well, wasn't that the aim of the exercise? thought the acoustician!).

It turned out that one of the visiting technicians had uttered a disparaging remark while on stage, and the star singer who was resting in the audience area had heard him and gone ballistic! Needless to say, nothing further was heard regarding the supposed acoustic defects.

13.6.4 Auditorium Maurice Ravel in Lyon

The Maurice Ravel Auditorium was built in the 1970s in the city of Lyon, France. While it can claim to be one of the first buildings using pretensioned concrete techniques in France, funnily enough, there apparently was no real acoustic study of such a facility that was intended for music performances. This eventually resulted in a rather too low volume for symphonic purposes, with a reflective concave back wall. More to the point, the stage had no lateral walls, and it was rather hard for the musicians to properly hear each other. The back wall had to be absorptively treated later on in order to prevent the occurrence of strong echoes, thus further reducing the reverberation time value. This eventually led to the use of an acoustic enhancement system by Philips in order to achieve a better sound coverage of the stage and audience area, as well as a greater reverberation time value.

13.6.5 ToneHalle in Düsseldorf

The ToneHalle in Düsseldorf is actually a former planetarium that was turned into a concert hall facility due to the historical tradition of the town enjoying a concert hall at this location. Due to its very concave shape of the envelope, there were quite strong echoes. When a major renovation of this facility was undertaken, the envelope was treated as very absorptive. In order to enhance the envelopment and increase the reverberation time, an acoustic enhancement system was successfully implemented [10, 11].

13.6.6 Royal Albert Hall in London

Prior to its renovation, it was said that the Royal Albert Hall was a place to play the most obscure works, as they would then have a chance to be heard twice. This was due to the very strong echoes occurring in this concave facility.

The acoustic study called for scale model investigations that helped devise the nature of diffusive and absorptive treatments to be applied.

13.6.7 A Small Music Room

A small (capacity 20 people) music room was built in a college. Right from the start, the staff wanted a chance to enjoy variable acoustics in it. This was achieved using curtains on the walls, as well as light absorptive panels that could easily be handled by small students and hooked on the walls. The reverberation time value could range between 1.8 and 1.2 s.

Lesson Learned: Variable acoustics can be pretty simple as long as you have the manpower (and the knowledge) to operate them.

13.6.8 The Best Concert Hall Ever?

A famous conductor was asked by a reporter: "Please, master, tell us, which is the best concert hall in the world?" Back came the answer: "What a question, it obviously is the one where I got the most applause!"

Lesson Learned: There really are many factors in the appreciation of a facility!

13.6.9 A Misplaced Lighting Fixture

When it was decided that the local concert hall facility was to be thoroughly refurbished, the mayor of a large town went visiting a few concert halls around the world. When he eventually visited the Sydney Opera House, he was so much impressed by the main lighting fixture that he decided he would have a similar one in his town.

Of course, the Sydney lighting fixture was meant to provide diffusion and was adapted to the acoustics of the hall. This was far from being the case in the unfortunate mayor's place, and its presence deteriorated the acoustics to such a point that the singers and musicians threatened to boycott the place as long as it remained in place, which led to its being dismantled.

Lesson Learned: Do not introduce changes in the design without knowing the possible consequences.

13.6.10 La Halle aux Grains in Toulouse

La Halle aux Grains (The Grain Market Hall) in Toulouse initially had been the market place for cereals since 1864. It was eventually turned into a sport hall in 1952. At that time, nobody cared much about the traffic noise from the two nearby streets. However, in 1974 work in the regular symphony hall prompted the Toulouse symphonic orchestra to move temporarily to those premises. Conductor Michel Plasson fell in love with the place. The audience was quite pleased too, as due to the circular shape, one could even be seated behind the orchestra. However, everybody concurred that traffic noise intruding inside the facility was way too high due to the large and leaky windows. Eventually, in 1988, serious rehabilitation work was performed, and this hall, seating 2200, now is the regular place for the symphonic orchestra.

REFERENCES

1. L. Beranek, *Concert halls and opera houses: Music, acoustics, and architecture*, 2nd ed., Springer, New York, 2004.

2. M. Asselineau, Du ballet au banquet en baskets (From ballet to banquet through basketball, in French), presented at ICA95 Proceedings, Trondheim, 1995.
3. M. Asselineau, *Quelques éléments d'histoire de l'acoustique* (*A few elements on the history of acoustics*, in French), Paris, 1984.
4. Décret n° 98-1143 du 15 décembre 1998 relatif aux prescriptions applicables aux établissements ou locaux recevant du public et diffusant à titre habituel de la musique amplifiée, à l'exclusion des salles dont l'activité est réservée à l'enseignement de la musique et de la danse (French decree about prescriptions applicable to public receiving facilities and rooms using amplified music with the exception of dance and music teaching reserved rooms), *Journal Officiel de la République Française*, December 16, 1998.
5. *Cahier des charges des salles Zenith* (*Specification book regarding Zenith type halls*, in French), 4th ed., COKER Company, Paris, 2002.
6. M. Asselineau, Acoustique active et acoustique passive—Approches complémentaires ou opposées? (Aactive and passive acoustics—Complementary or opposed ways? in French), presented at CFA2010 Proceedings, Lyon, 2010.
7. B. Shield, T. Cox, Concert hall acoustics tutorial, Salford University, Manchester, 2000, http://www.acoustics.salford.ac.uk/acoustics_info/concert_hall_acoustics/?content=shape.
8. M. Barron, C.B. Chinoy, 1:50 scale acoustic models for objective testing of auditoria, *Applied Acoustics*, 12, 361–375 (1979).
9. V.M.A. Peutz, B. Bernfeld, Variable acoustics of the IRCAM concert hall in Paris, presented at Proceedings of the 10th ICA, Sidney, 1980, Paper El.3.
10. Peutz, http://www.peutz.de/pdf/tonhalle_144.pdf.
11. Peutz, *Acoustics by Peutz—Theatres and concert halls*, The Hague, 2010.
12. *J. Attali, Bruits—Essai sur l'économie politique de la musique* (*Noise—Essay on political economy of music*, in French), Fayard, Paris, 1977.
13. Peutz, Standard guide for initial design, Internal document, 2004.
14. A. Gade, *Acoustics in halls for speech and music*, Springer Verlag, Berlin, 2007.
15. M. Long, *Architectural acoustics*, Elsevier Academic Press, Burlington, MA, 2006.

Chapter 14

Operas

14.1 INTRODUCTION

Operas are an interesting mixture of theatre (due to the play on the stage) and concert (due to the orchestra). Inevitably, one will find features of both theatres and concert halls; no less inevitably, there will be compromises to settle.

Opera facilities are meant to provide the audience with a good view of the stage and a good hearing of the singers on stage and of the musicians; in addition, they must enable the musicians and singers to hear each other.

14.2 REQUIREMENTS

14.2.1 Foreword

An opera house is, of course, a building where the opera hall and its stage are housed. But there are quite a lot of other spaces (e.g., rehearsal rooms, plant rooms, technical rooms, etc.) that have their own requirements.

As with any project, there usually exists a program: What does the end user (and the payer too!) actually want? This means the various requirements must be identified and the relevant acoustic objectives stated. In today's spirit of sustainable development, one must be conscious that a team effort (i.e., architects, structural engineers, HVAC, and acoustics, just to name a few) occurs and make sure that the various solutions that are considered at the design stage are compatible with each other. Acoustics is often part of a global problem that can only be solved by a complete design team fed the relevant data by the end user. This is especially true in the opera projects, as many features will turn out to be the result of compromises between the architect, the acoustician, and the stage engineer, just to name a few. In addition, due to the cost of such a project, the city or region paying for it will certainly have some requirements regarding the outward appearance of the building, as well as some internal features.

14.2.2 Opera Hall

To start with, there is a visibility requirement, as the stage must be properly observed by the audience. This usually induces a slightly stepped audience seating similar to that of a theatre.

Next, there is a reverberation requirement. This will seriously depend on the priority awarded to either the music or the sung text. This choice is not a new problem at all: In the 18th century, there had been quite a feud, known as the "bouffons' quarrel," where the French composer (and French music proponent) Rameau advocated proper understanding of the text, while the French composer and writer (as well as Italian music proponent)

Rousseau pushed for proper musical quality over speech intelligibility [1]! More recently, it was suggested that in America opera is seldom sung in local language, and people simply come to hear the music [2]. This eventually results in a broad array of reverberation time (RT) values according to the nature of the plays: For example, the Scala of Milano features a 1.1 s RT in order to enhance the vocal articulation performance of the singers in the Bel Canto works, while Bayreuth features a 1.8 s RT compatible with the symphonic music of Wagnerian works. The latter provides a perfect example of blended skills, as Wagner designed this facility in order to be able to produce his works, while his works were written with the capabilities of Bayreuth in mind. It gives a proper balance between a full orchestra of 100 to 130 musicians and the singers due to a covered pit [2].

Much debate is to be expected from the orchestra pit. In the early days of opera, it was not unusual for the orchestra to be split and located on stage on either side of the playing area. With the development of stage techniques, the orchestra found itself relegated at the foot of the stage and slightly under it. This feature was taken advantage of to try to partially cover the orchestra to get a better equilibrium between the radiated sound power level of the singers and that of the orchestra. But the move was not really popular with many conductors, as their beloved orchestra would be largely hidden from the audience!

Putting everything into perspective, it is perhaps not that surprising that a sizable number of seats are not considered fit for opera, event in famous halls [3]!

14.2.3 Other Rooms

Here there is no real visibility requirement (unless rehearsals are opened to the public). But there undoubtedly are acoustic requirements.

The reverberation requirement will depend on the actual use of the room. Usually professionals will prefer a rather dry room (compared to the stage acoustics) that will enable the conductor or the choir master to spot any defect. One will often end up with a RT ranging from 1.0 to 1.5 s, with the higher values usually given to choir rehearsal rooms.

Sound insulation with regard to other spaces must, of course, be properly carried out.

14.3 ACOUSTIC TARGETS

In the olden days, the sound insulation requirements with regard to the outside were not that high, as there were no sound systems except the symphonic orchestra. Actually, it could be nil when the performance was given on the terrace in the gardens of a manor, or on the stairs of the church (as was common for oratorios).

Gradually, sound systems were introduced and the frequency domain of interest was steadily enlarged, while performances found themselves under a roof. But even then, some facilities had been tackling the low-frequency range: an organ with a 32 ft tube can generate sounds down to 16 Hz, which is useful as a fundamental for creating sounds, and those were already in existence in the 19th century.

Nowadays, modern sound systems can generate high noise levels over a very broad spectrum. This means that the sound insulation with regard to the outside must be treated accordingly. There usually are some legal requirements (e.g., in France, a decree on musical venues using amplified music). In addition, care must be taken regarding the protection of the hearing of both the audience and the musicians (there is work pending in this matter in the European Union). Compared to a regular concert stage, the orchestra pit is usually a rather cramped affair, especially in old operas (e.g., the Capitole in Toulouse), and sometimes there will be a choice to make between an absorptive ceiling (that will reduce the radiated sound power level of the orchestra, but nevertheless save the eardrums of those close to the tubas!) and a reflective one.

As usual, the fundamental questions will be: What does the end user want to do, and what will the project actually allow him? Unsurprisingly, the acoustics of an opera hall will combine those of a theatre (for speech and visibility purposes) and those of a concert hall. Inevitably, one will have to choose sides: speech or music oriented?

The internal acoustics of an opera facility must be adjusted to its finality: When speech oriented (or dealing with modern music generated through electroacoustical means), the basic idea is to avoid acoustic reflections on the walls and reduce the reflections on the ceiling (as much as in a theatre), which leads to a rather low RT due to the absorptive treatment of the opera hall. On the opposite end, when music oriented, one looks for an enveloping sensation through numerous reflections and diffusion on the walls and ceiling (meaning acoustic reflections on the side walls must be provided, and acoustic reflections on the ceiling should be provided too). Sound propagation from the stage and pit to the farthest seats should be ensured. This means that a significant part of the ceiling should be reflective and oriented so as to ensure correct sound distribution over the audience.

Looking at the internal acoustics of opera halls, one must perceive the sound as coming from the place on the stage where the instruments are (and even from backstage, as some composers require such an effect too); even then, one may distinguish between the pure classical works, where each set of instruments of the orchestra must clearly be distinguished, and the later works, where the orchestra is treated as a global mass.

Eventually, the rule of thumb makes one head for a reverberation time in the range of 1.3 to 1.7 s and an early decay time (EDT) value close to the RT value, while the clarity C_{80} is to be found in the range of 0 to 3 dB [8].

The sound levels generated by mechanical equipment should not be a hindrance. This typically means a background noise level value in the 25 to 35 dB(A) range.

14.4 A FEW BASIC RULES

As with any project, an opera project starts with a program: What does the end user (and the payer too!) actually want (e.g., capacity, shape of the hall, type of performance)? This means the various requirements must be identified and the relevant acoustic objectives stated. In today's spirit of sustainable development, one must be conscious that a team effort (i.e., architects, structural engineers, HVAC, and acoustics, just to name a few) occurs and make sure that the various solutions that are considered at the design stage are compatible with each other. Acoustics is often part of a global problem that can only be solved by a complete design team fed the relevant data by the end user.

Assuming that the emphasis on the facility really is opera, the first point will be to define its capacity. This is no academic question, as an opera is staff-consuming and one has to pay their salaries: In addition to the musician and the singers, one needs the electricians and machinists to work the various devices and handle the back stage and gratings. In order to balance the finances, this means that a large audience must be accommodated. In order to achieve such a feat, balconies will be needed. More to the point, according to the size and shape of the orchestra pit (which typically occupies what would be the four front rows), the capacity of the audience may be limited.

In order to increase capacity, it might be tempting to increase both the height (to add one more balcony) and the length (to add more rows). However, the former move will increase the RT lower limit, while the latter will boost the chances to have an echo appear, while decreasing the stage visibility of the audience. More to the point, the local urban planning rules may not take too kindly to a higher-than-usual building being erected, so it will be necessary to dig, and apart from the usual construction problems, this will also double the needs for safety exits according to most safety rules.

Then there will be a few choices to make regarding the stage size and equipment: Does one want a swiveling double stage so as to be able to change the scene in a hurry? Does one need some stage machinery to elevate or lower part of the stage floor? How much side space will be needed? The answer to those questions will have an impact on both the required surface and the acoustic noise control measures to be implemented.

Now, what acoustics are looked for? The answer will be complex, as it depends on the wishes of the conductor (who might prefer an open pit as a better showcase to his orchestra), the physical constraints of the facility due to the available volume and shape, and eventually the doctrine of the local cultural authorities: Does one favor the speech intelligibility or the music fullness? Usually, one will end up with absorptive boxes (in order to avoid unwanted coupled echoes) and diffusive treatment over the front of the balconies, while the walls will be treated with either diffusive or absorptive coverings in order to suit the reverberation time targets. The ceiling will be partly reflective, and its shape will be designed so as to help propagate the sound energy from the stage and pit areas to the audience. The stage tower will be treated to a shock-resistant absorptive material.

Eventually, one will usually end up with a volume in the 6000 to 10,000 m³ range [8].

There is no question about the air handling in the facility. The best way to cool the audience is to insufflate air at low speed under the seats. But care must also be taken for a proper cooling and ventilation of the orchestra pit due to the amount of active people and the lighting fixtures. Last but not least, the stage must be ventilated and cooled too. The problem is to achieve this without having the back scenes moving due to the air speed, which calls for sizable louvers to achieve the required low speed and large flow rate.

Are we done now? Not yet, as the performance hall is only a small part of the whole facility. In order to operate smoothly, one needs rehearsal rooms, at least one for the choirs and one for the orchestra, not to mention the various groups of instruments. While there probably will not be a performance held during the rehearsals, all those spaces must be able to operate simultaneously. This usually requires highly insulated rooms built as a box in box. One must also take into account the need for consequent height and volume (e.g., both the choir's and the orchestra's rooms will be double height compared to other spaces).

Last but not least, there will probably be ballet included in the activities of the opera house. This means that a dance rehearsal and training room must be provided. Keeping in mind that a minimum 4.5 m clearance is required for such a room and that 20 people jumping together provides quite a lot of excitation for the structure, strong impact noise control measures will be needed. In addition, the dancers are—understandably—particular on the kind of flooring they want, so this must be cleared well in advance during the design process.

14.5 MODELING

14.5.1 Foreword

Modeling is a nice tool, both for the specialist and for public information. Like all tools, it does have limitations and cannot be trusted blindly! While it is nice to establish a comparison between different situations—provided the model has been properly elaborated—it seldom comes up with an absolute value to be trusted and cannot be operated by nonspecialists. Here the problem is especially complicated by the numerous volumes (stage, hall, orchestra pit), as well as by the somewhat elaborate decorations used for the inner surfaces of the facility.

14.5.2 Ray Tracing Modeling

Ray tracing using a computer does look the part. However, the physically minded can quickly understand that solving the wave equation inside a hall full of singularities, such as balconies, coupled volumes (e.g., orchestra pit, back stage), and funny-looking surfaces (including curved surfaces and complex decorated walls), is not going to be easy, not to say accurate. Nevertheless, ray tracing is rather quick to implement (especially if the architect has previously been talked into providing the relevant drawings suitably clean of unnecessary details) and can be used to pinpoint and correct such defects as echoes and inadequate reverberation time.

Some commercial programs now provide auralization modules that enable the user to hear how the hall under study will eventually sound. This may come in handy when it comes to the discussion of the balance between the orchestra and singers.

14.5.3 Scale Modeling

Scale modeling may look old fashioned, but it always features a kind of magic to onlookers. Actually, one may discuss scattering objects for hours in front of a dubious client using some computer simulation results and images without much success, but having the client look at a scale model while being told the problem is usually quite effective, not to say awesome.

Scale modeling proves efficient when dealing with singularities such as balconies, coupled volumes, and funny-looking surfaces.

What scale should be used? Clearly enough, the smaller the scale, the higher the frequency range will be and the more complicated the measurements will turn out, as air absorption becomes quite a problem. Scales like 1/10 to 1/16 are rather nice and popular, as air absorption is not yet too much of a problem, but while the model is easy to work with, it also happens to be quite cumbersome. Experiments have been performed with cardboard 1/50 models [4].

14.6 EXAMPLES

14.6.1 The Theatre of Mirecourt

Mirecourt is a pretty small eastern French town. It features a small theatre-opera facility seating 200. The unknowing would wonder why such a small town would have such a facility until discovering that there are many famous string instrument manufacturers living there. It featured a reverberation time of 1.5 s, but the stage area and open orchestra pit area were quite limited. When a possible major renovation of the facility was studied at the end of the 1990s, it was found that an increase of capacity would not be possible without significantly altering the architectural appearance of the place. As the operation was uneconomical, it was decided to drop the opera and theatre activities.

14.6.2 The Opera of Lyon

For many years the French city of Lyon had enjoyed a facility known as the grand theatre. It featured a performance hall and its entrance hall, as well as a small stage tower and administrative or technical rooms. The presence of an orchestra pit offered the opportunity to hold opera performances too.

In the 1980s, an ambitious project was launched to provide Lyon with a real opera facility. An international competition was held; due to the historical interest of the building, the façades and main foyer were to be kept. Architect Jean Nouvel won this competition and designed a construction in which the lower part was made of the existing façades. It was topped by a large construction terminating with a glazed half cylinder that was housing the administration, the dressing rooms, and the dance rehearsal and training rooms.

Due to the limited surface of the allocated area, extra space had to be found vertically. This resulted in a configuration where the choir's and the orchestra rehearsal rooms were located in the basement, with their attending smaller individual spaces located around.

In order to achieve the loading/unloading of the back scenes, a larger door than the original was needed. However, due to the historical classification of the building's façades, it was out of the question to open a new door. It was eventually found possible to have part of the façade as a mobile unit with an acoustic door behind, with space enough in the elevator to hoist a full trailer.

The opera itself seats 1300 with three balconies; with a reverberation time of 1.3 s, it is particularly well suited for French opera, as it is a good compromise between intelligibility and musical fullness. In order to increase the sound insulation of the performance hall with regard to the rehearsal rooms and the street level, it is suspended from the upper part of the structure [5].

14.6.3 The Opera of La Bastille

When it was decided that the city of Paris needed a new opera facility, the site of a former railway station place, de la Bastille, was retained. In order to answer efficiently the needs of modern productions and serve the cultural and teaching requirements, it features a 21,000 m³ main performance hall seating 2700 with a reverberation time of 1.7 s, with a smaller hall for concerts and small lyrical works, which doubles as a conference facility. In order to reach the required capacity while staying at a reasonable sight distance from the stage, it features two frontal balconies. The stage is fitted with a large turntable, which allows the back scenes of an act to be prepared while the previous one is underway.

14.6.4 The Opera of Vichy

Vichy used to be a central France town well known for the properties of its waters. This attracted many tourists, and in the 1850s the French emperor Napoleon III made the place fashionable. A casino and a small concert hall were built to accommodate the leisure time of the visitors, with plans to build an opera. The latter was eventually commissioned in the 1930s and was considered a reduced copy of the famed Wien Opera. It seated 1500 on the ground floor and two balconies. It can be virtually visited in [6]. The art deco facility featured numerous wall paintings on the nonabsorptive walls. The potential risk of echoes due to the horseshoe shape of the hall was somewhat tempered by the presence of numerous boxes. But there also was a curved proscenium ceiling and a cupola over the audience that gave very strong reflections, to the extent that on the first balcony spectators would turn back their head to the right of the hall when a soloist singer would act on the left of the stage. An acoustic diagnosis performed in the mid-1990s [7] actually showed that at the first balcony, there was more sound energy coming by reflections than directly.

Even with those defects, the opera has been enjoying good attendance. For many years, the orchestra and singers of the Paris Opera would have a short summer season in Vichy, and nowadays, the facility still enjoys good attendance in the summertime.

14.6.5 The Theatre-Opera Capitole of Toulouse

The Capitole is the historical theatre-opera of the French city of Toulouse. The building was built in the 17th century to house the town hall and the opera. This opera now houses a yearly bel canto competition and has long been noted for a tough exigent audience. Over its existence, it has lived through various accidents and fires that were followed by partial reconstructions. The audience part was refurbished in the 1990s, and the stage tower, which featured manually operated old block and tackle systems, was earmarked for later. Following the accident of the AZF chemical plant in 2003 that damaged roofs around the city, it was decided to perform a major renovation of the stage tower, which had structurally suffered. This opportunity was taken to also perform a significant renovation of the orchestra pit to try to enlarge it.

The design was a bit hectic, with limited documents available. When the work started (by uncladding some structures and making tentative holes in the walls), it was found that most of the dimensions on the existing drawings were wrong. The first step was for the structural engineer to understand how the structure was supporting itself, and for the architect to try to squeeze in whatever rooms were needed. Prior to the (somewhat destructive) soundings made by the structural engineer, the acoustician made an acoustic diagnosis of the existing facility and interrogated the musicians and technicians about their eventual wishes and complaints. A new floor was built for the stage, and an elevating platform was included. The stage tower was heightened (which was no small feat, as this needed a special building permit from the historical monuments department; in addition, it was necessary to introduce extra foundations to cope with the extra weight, and the subway tunnel was not far from the building). This enabled the creation of a real grill over the whole stage, as well as the creation of a motor room for all the hoists.

At the request of the technicians, the openable windows were kept, but a system of double windows was introduced to keep away the noise from the busy streets nearby.

The orchestra pit was slightly enlarged and acoustic treatment was basically diffusive, with the possibility of using stage-like curtains if some deadening of the sound was wished for. Unfortunately, it was not possible to really increase the height of the pit under the stage floor.

Ventilation and cooling of the premises proved to be a real challenge in a building that was not initially intended for such a treatment. Space had to be found in the attic and in some passageways, while a few rooms had to be downsized to allow for the routing of ductwork.

REFERENCES

1. J.-J. Rousseau, Maintenant que les bouffons sont congédiés, ou prêts à l'être (Letter on French music, in French), 1753, http://www.musebaroque.fr/Documents/rousseau_lettre.htm.
2. L. Beranek, *Concert halls and opera houses: Music, acoustics, and architecture*, 2nd ed., Springer, New York, 2004.
3. G.C. Izenour, *Theater design*, McGraw Hill, New York, 1977.
4. M. Barron, C.B. Chinoy, 1:50 scale acoustic models for objective testing of auditoria, *Applied Acoustics*, 12, 361–375 (1979).
5. Peutz, Lyon opera, measurement and concept reports (in French), 1991.
6. http://www.ville-vichy.fr/Visite-virtuelle-Opera-de-Vichy.
7. Peutz, Vichy opera, initial measurement report (in French), 1993.
8. Peutz, Standard guide for initial design, Internal document, 2004.

Chapter 15

Multipurpose Facilities

15.1 INTRODUCTION

As implied by their denomination, multipurpose facilities are meant to house practically everything! Therefore, they must be flexible venues easily adapted to the required use.

15.2 REQUIREMENTS

Everything should be possible; at least that is the idea of the end user. More to the point, this probably is the idea of the future operator too (even if he has not yet fully expressed it to the end user). One must not treat this preliminary aspect lightly, as all of the project will hinge on it.

As with any project, there usually exists a program: What does the end user (and the payer too!) actually want? This means the various requirements must be identified and the relevant acoustic objectives stated. In today's spirit of sustainable development, one must be conscious that a team effort (i.e., architects, structural engineers, HVAC, and acoustics, just to name a few) occurs and make sure that the various solutions that are considered at the design stage are compatible with each other while staying compatible with the user's requirements. Acoustics is often part of a global problem that can only be solved by a complete design team fed the relevant data by the end user.

While quite a number of activities can usually be accommodated under the same roof, caution must be exercised: Basically, for any activities save for classical music, one can use a rather normal space, that is, as long as the sound insulation requirements with regards to other spaces (including the outside environment) are met, together with the internal acoustic requirements (mainly reverberation control and mechanical noise control). One will then probably end up with a rather absorptive inner envelope (to provide reverberation and noise control). But should classical music be considered, then this is an entirely different story, as acoustic reflections will have to be provided.

15.3 ACOUSTIC TARGETS

The acoustic targets should of course be adapted to whatever use is aimed for.

This means that one may either set an average given reverberation time (RT) value supposed to be suitable to all kinds of uses, or manage to have variable acoustics.

Modern sound systems can generate high noise levels over a very broad spectrum. This means that the sound insulation with regards to the outside must be treated accordingly. There usually are some legal requirements (e.g., in France, a decree on musical venues using amplified music). In addition, care must be taken regarding the protection of the hearing of both the audience and the musicians or technicians.

Therefore, the fundamental questions will be: What does the end user really want to do, and what will the project actually allow him? It will have to be addressed during all of the project duration, as some compromises have to be reached (e.g., providing heavy lifting capacity under the roof structure while keeping an insulating ceiling, using the thermal inertia of the building while providing acoustic absorption, etc.).

The sound levels generated by mechanical equipment should not be a hindrance; yet it is also useful as a masking noise. This typically means a background noise level value in the 40 to 50 dB(A) range inside the premises. It may actually be lower should a higher quality be aimed for (but then somebody had better make sure that proper care has been given to correct ventilation and cooling).

15.4 A FEW BASIC RULES

As with any project, a multipurpose hall project starts with a program: What does the end user (and the payer too!) actually want? Is he ready to sacrifice a few possibilities in order to achieve better results in a given use, or simply to save money? This means the various requirements must be identified and the relevant acoustic objectives stated. Should some prescriptions be definitely ruled out, somebody has to explain to the end user the advantages and disadvantages of such a move. As stated in Section 15.2, acoustics is part of a global problem that can only be solved by a complete design team fed the relevant data by the end user. One should also think of maintenance.

One of the first questions to be addressed is: Does one want to be able to operate using a normal voice? Should the answer be affirmative, this will mean a strong limitation of the dimensions of the hall (a distance of 20 m from the stage to the listener will typically be considered). Next, one should consider whether classical music is intended or not. Should that be the case, this is a real problem, as lateral reflections will be needed for a good musical sound quality, while it will have to be excluded for speech intelligibility! To put it bluntly, one will then be left with a choice of designing a concert hall that might accommodate some other activities or making a theatre/cinema/exhibition hall that might accommodate some classical music [1]!

Do beware of operators stating that they will never use amplified music! It is inevitable that in a multipurpose venue amplified music will be used. Such an answer means that either the operator is not conscious of the risks involved or he is knowingly avoiding the issue.

Due to the broad range of activities likely to be pursued in a multipurpose hall, variable acoustics are usually required. Those are achieved either through mechanical systems changing the appearance of the walls and ceiling, and even the volume of the hall, using mobile walls and ceilings, or by means of electroacoustic enhancement systems.

Mechanical systems typically call for mobile panels on the walls and ceilings, featuring at least an absorptive side and a reflective side. Depending on the number of panels of a given absorptive coefficient, a more or less suitably high RT value can be achieved. In addition, parts of the ceiling can be lowered in order to reduce the mean free path, while parts of the walls can be moved closer to the stage if needed. This is quite logical, and it often appeals to the end user, as it calls for the various trades of the building industry. Now, such systems require maintenance, and unless they are motorized, they also require some suitably educated staff to operate them.

To start with, electroacoustic systems are typically implemented in as absorptive as possible a hall, as they are primarily meant to introduce extra reverberated energy. They require highly specialized personnel. Basically, they are made of a number of loudspeakers along the ceiling and walls that broadcast the kind of signal that would be expected from a natural reflection on a physical surface. Of course, the elaboration of this signal through signal processing units takes into account the distance from the source to the audience. This means they can be used not only for mere reverberation purposes, but also to create extra lateral reflections to enhance the envelopment. Such systems require a period of tuning and can be quite complicated if there are several configurations; on the other hand, they usually can be reprogrammed to cope with new situations (though it does take time). A word of advice here: Such a system can add energy but not suppress it, so if the RT was already too long or if there was a serious defect such as an echo in the hall, the problem will not be solved. One must also remember that such systems will work from the sound in the hall—any sound, that is. It means that should the HVAC system be a bit too noisy or the microphones of the system be implemented a bit too close to an air outlet, the corresponding noise will be amplified and distributed over the hall too.

Unloading and loading of supplies for an event can be quite noisy, and specific precautions are needed if one wants to keep peace with the neighborhood. This often means that covered facilities (e.g., a loading platform) are needed to prevent the noise of such activities to be radiated around, starting with the backing sound signal of vehicles.

15.5 A FEW TYPES

15.5.1 Foreword

Multipurpose venue will often mean *variable acoustics venue*. Unfortunately, the latter implies the presence of a staff capable of operating the corresponding devices; experience often shows that the available personnel do not systematically have the competence or the inclination to spend time operating those devices.

15.5.2 Light Multipurpose Hall

As implied by its name, the light multipurpose hall is supposed to be a hall where one can perform various kinds of activities without resorting to heavy means to change the acoustics of the hall or its morphology. This often is the case for multipurpose venues of small townships. There typically will be a small stage at one end of the hall, with no possibility of relocation.

This means that either a mean RT value acceptable for the various activities to be carried out is chosen, or a slight adjustment is made possible using reflective walls in front of which heavy curtains can be drawn to add extra absorption when needed. Typically, such a system will be used with reflective walls for instrumental music purposes and with absorptive walls for speech purposes.

Typically, with a fixed value, the RT will be found in the 1.0 to 1.3 s range. Using curtains on a large scale, it often is possible to reduce those values by 0.2 s.

15.5.3 Heavy Multipurpose Hall

In such a facility variable acoustics will be used, and often there will be some significant stage equipment. It is also not uncommon for such halls to have the possibility of relocating the stage (e.g., frontal by the small or the large wall, or central). Variable acoustics are used to try to adapt to the intended use, either through mechanical systems changing the appearance of the walls and ceiling, and even the volume of the hall, or by means of electro-acoustic enhancement systems. Quite often, there also are retractable stepped seats.

15.5.3.1 Variable Acoustics through Mechanical Means

The envelope of the hall will often feature dual-sided panels that can be rotated, with one face absorptive and the other reflective. Quite often, this is associated with a variable height ceiling, with a typical 8 m high ceiling for theatre purposes and 15 m high ceiling for music purposes. More to the point, extra-reflective panels can be brought in close to the musicians in order to introduce some early reflections. Last but not least, should music and theatre be considered, a mobile proscenium wall will be needed. Such a wall typically features an upper part that can be lifted inside the stage tower and the lower parts (on each side of the stage opening) that swivel onto the structural walls so as to be able to choose between a small opening with a proscenium wall (theatre mode, with two distinct volumes for stage and audience, respectively) or a large opening (music mode with stage and audience within the same volume) [2, 3].

15.5.3.2 Variable Acoustics through Electroacoustic Enhancement Systems

There now are a few electroacoustic enhancement systems available on the market. According to the degree of enhancement needed and the multiplicity of situations to be coped with, their complexity is more or less important. They can manage to turn an initially rather dead hall (e.g., with a RT of 1.5 s) into a lively space with a RT greater than 2.2 s. [4]

15.5.4 Sports Hall

Nowadays, most sport halls are also used for other kinds of performances; for example, they may house a basketball competition on one day and a pop concert the next. This is a low-cost approach for facilities of circa 4000 seats, and in order to keep the cost down, there usually are no fancy variable acoustics to be found in there.

Due to the sports activity, the absorptive treatment is shock resistant. In order to satisfy the pop music requirements, it is also quite efficient in the low-frequency range. One eventually typically ends up with a U-shaped seating audience, with the front end accommodating a removable stage. Experience shows this works rather well, with a typical RT value of 1.5 s. In some more elaborate venues, the stage can actually be located anywhere on the sport field and extra retractable tiered seats can increase the capacity when needed (i.e., when not used in sport mode).

There is a bit of difference when it comes to noise levels generated: Pop venues can generate much more low-frequency noise than sport, and the sound insulation of the hall with regards to the exterior space will have to be treated accordingly. Another issue is the noise from HVAC: While a 40 to 45 dB(A) noise level can still be admitted in a regular sport hall, it is way too much for a performance hall; proper noise control measures (meaning low speed air supply and intake) must be thought of.

15.5.5 Exhibition Hall

Due to their large size, exhibition halls can be used for practically everything, from various size exhibitions to banqueting through performance, conference, and examinations. In order to be able to fulfill those various roles, there are basic acoustic requirements to be satisfied.

To start with, a proper sound insulation with regards to the exterior space and the adjoining spaces (remember, a hall can be allocated to a specific task, while the next hall can either be under reconfiguration or allocated to another job, e.g., exhibition) must be achieved. This is easier said than done, for most of the time the separating walls to adjoining spaces are made of mobile walls. The best of those mobile walls feature a sound reduction index of over 50 dB when tested in the laboratory. In practice, due to their frequent use (once per day is not uncommon), the seals are quickly worn out and their performance can get down to 35 dB over 10 years [5]. There often are large doors to the outside for both safety purposes and the easy loading of exhibitions items, and it may be necessary for both thermal and acoustical purposes to provide an airlock (please note that the firemen may be reluctant, so this must be discussed as early as possible in the design process).

The RT should be kept rather low (i.e., in the 1.0 to 1.3 s range). More to the point, the spatial sound level decay should not be smaller than 3 dB(A) per doubling of distance in the empty facility.

Finally, the sound levels generated by the mechanical equipment should be kept in a suitable range: It must not be too high for performance and conference purposes, as well as banqueting, but it must be high enough to be of some use as a masking noise. This typically calls for a 40 dB(A) value, which is suitable for a conference with a sound system.

15.5.6 Studios

Studios are primarily meant to record sound and images. This means that the sound insulation with regards to other spaces must be particularly specified. More to the point, in order to perform a good recording, the acoustic signal must not be perturbed by undue reflections; this means that absorption must be applied on the envelope of the room. In order to ease the work of the sound engineer, the basic idea is to try to get as flat as possible a reverberation spectrum. Values in the range of 0.2 to 0.6 s are common for such spaces. Finally, good noise control measures must be taken against the HVAC noise. This means that significant silencers and very large sections of ducts must be used.

It is not uncommon for studios to be also used as a small performance space. The tiered seating must not generate noise (e.g., a stiff structure, not squeaking, with a carper-style floor covering).

15.5.7 Outdoor Extensions

While outdoor extensions are not considered true buildings, they often are found next to a built facility and need addressing. For example, some indoor performance facilities feature an outdoor tiered area at the back, so as to be able to use the stage equipment. Also, most exhibition halls feature outdoor areas complete with electric and fluid supply points in order to boost the number of exhibitors, provide animation, and give a chance to be able to exhibit large items.

Unfortunately, there is noise associated to such facilities: First, preparing and later dismantling the exhibition calls for delivery trucks and forklifts, with their usual honking and clanging noise. As the operators usually prefer to perform such activities at nighttime, this is a potential source of community noise to be dealt with (usually through the use of suitable noise barriers around and proper instructions given to the crews, as well as specifications regarding limits set on the duration of the event). Next, during normal operations, in addition to the noise from the crowd, there usually is noise from a sound system. One must be aware that such noise must comply with the regulations in force. There may be a few specific authorizations issued for specific events with higher admissible noise levels, but those are limited in time and scope. The basic trick is to try to use a high number of rather directive loudspeakers. This way, the attendance is always in the direct field of a loudspeaker whose sound power level can be adjusted. They must be installed low enough so as to benefit from the noise barriers around the site.

For large outdoor venues, it is even possible to recreate an impression of ceiling and lateral reflections through the use of dedicated electroacoustic systems [4].

15.6 REHABILITATION

Rehabilitation is a quite usual procedure in Europe [6]. There may be quite a few reasons to go for it. To start with, older buildings usually feature a higher potential occupation ratio per ground square meter than the one allowed by normal recent constructions. More to the point, it helps the urban planners to keep a homogeneous urban appearance. Last but not least, it helps save time if the walls and floors (and sometimes even the roof) are kept; administratively speaking, it may also speed things up, for when the envelope of the building is kept, one often will solely require a fitting out permit instead of a full building permit. When it comes to multipurpose facilities, rehabilitation can be a good opportunity to add a few extra capabilities to the facility.

In such a kind of project, it is necessary first to perform a diagnosis of the existing building in order to find out how it is built and where the sensitive points are. Each specialty will have to perform its own diagnosis. One should remember that a structural diagnosis usually can be a bit destructive, as the structural engineer will typically cut through floors and walls in order to assess their composition, so the acoustician must make his measurements first.

Note: There is a strong need for coordination between the interested parties, as depending on the planned sketch of the future rooms inside the building, some zones will be more sensible than others. More to the point, everybody must be aware of each other's needs.

In addition, one must be particularly aware of the safety requirements applicable: Whenever a derogation is delivered, it will certainly entail so-called compensatory measures that will probably be costly in terms of investment or extra personnel (cf. Section 6.10.8).

The diagnosis through measurements will feature:

- Sound insulation measurements between rooms when the walls are kept. Note: This measurement will help assess the potential flanking transmission by those walls, as well as the potential sound insulation of those rooms.
- Sound insulation measurements and impact sound measurements between rooms at different floors. Note: This measurement will help assess the potential sound insulation between floors.
- If applicable, vibration measurements on the floors and walls. Note: This will help assess the eventual noise generated by vibrations (e.g., from rail lines nearby) and vibration levels inside the building.
- An acoustic diagnosis of the site will be performed as per a regular new construction project (i.e., assessing the sound level values on the site and finding out what the potential noise sources around are, as well as the potentially nearby noise-sensitive zones). Note: Do keep in mind that usually a rehabilitation project will entail some demolition prior to the actual construction work. Under the nice words, one can already hear the concrete breakers hammering away, so one had better have a good look at the location of the nearest neighbors, especially those who are structurally linked to the building. It will probably be necessary to explain to the neighbors the basics of the project

and point out that while some phases of the work will be noisy, they will be kept to a minimum of duration and their time schedule will be adapted while appropriate noise reduction measures will be implemented.

It must be emphasized that the diagnosis will constitute the testimony to the acoustic performance of the building prior to any work. It is not only a basis for the acoustic studies (from which predictive computations will be elaborated), but it also is often a compulsory step to be able to ultimately prove that the initial acoustic performances of the building have not deteriorated [6].

In the particular case of historical buildings, things can get quite complicated, as usually the façades and even the roofs must be preserved. In some cases, it even is necessary to preserve some interior spaces (e.g., because of paintings on the walls or ceilings). Under such circumstances, the acoustic objectives must be adjusted on a case-by-case basis, and specific solutions must be developed (e.g., introducing intermediate spaces around in order to prevent direct transmission to other spaces of interest, or working on the other side of the partition or floor using a floating floor, a plasterboard ceiling, or half wall, with mineral wool in the void).

A special mention must be made regarding performance halls: Those usually are considered historical landmarks, and the end user may wish to preserve (or sometimes improve) much more than the sound insulation characteristics. This means that specific room acoustic measurements must be carried out in order to explain the physical phenomena that are behind the acoustic characteristics, and then questions must be addressed to the users to know for sure whether they actually want them to be kept as such. Only after that will it be possible to prescribe the relevant constructive solutions.

15.7 EXAMPLES

15.7.1 Vendespace

Vendespace is a multipurpose facility that is located close to the French city of La Roche sur Yon. It features a 4500-seat multipurpose hall, a sport hall, and a gymnasium, and was designed by architect Paul Chemetov.

In order to confer as much flexibility as possible to the multipurpose hall, which has already seen different events as a lyrical concert, a sport event, or a pop concert, part of the seating is made of retractable tiered seats. A grill covers most of the floor surface, enabling the installation of the stage nearly anywhere in the hall. The acoustic treatment calls for an absorptive material (mainly fibrous material protected by either a perforated metal panel or a perforated wooden panel). In order to get the acoustic conditions required by live music, the reverberant sound is provided by a reverberation enhancement system (Constellation by Meyer Sound), which helps recreate lateral reflections and increase the reverberation time value if need be [4].

15.7.2 Palio and Arena Loire

In 2004 a small township of southwestern France close to Périgueux had been enjoying a high-level basketball team. In order to give them a chance to play competitions at a high level, it was necessary for them to be able to welcome opposing teams in a minimum 4000-seat facility. While there was no regional budget for such a sport facility, some funding was available for a cultural facility. A quick survey showed the mayor that there was no such facility within 250 km, so there definitely was something possible in this field.

The requirement book by the mayor to the design teams was pretty simple: Design a facility that would seat a minimum of 4000 (for sport competition purposes) and no more than 4500 (for cultural events, as more staff would then be needed for safety purposes) and cost as little as possible! Local architect Bernard Chinours took the challenge and built Le Palio. This is a concrete structure (for the stepped floor) topped by metal walls doubled by a plasterboard half wall. The roof was a standard commercial type featuring a sound reduction index of 55 dB. The acoustic treatment called for absorptive fibrous material protected by a perforated metal plate. The HVAC called for two textile ducts suspended over the gratings. In addition to the usual team changing rooms, there were a couple of private dressing rooms too. Since its commissioning, the facility has proven its worth by housing sports events as well as pop concerts and even choirs!

In the same spirit, a small French town close to Angers chose to build the same kind of facility. Compared to the original, this one, named Arena Loire, features stronger low-frequency absorption so as to please the visiting sound engineers. During the inaugurating week, singer Barbara Heindricks declared herself satisfied with the acoustics of the place.

15.7.3 Paris Expo

The exhibition facility Parc des Expositions of Paris features several divisible halls. While those halls are primarily designed for exhibitions, they can also be called to serve other functions, such as large conferences, events, or banquets.

The older hall was demolished in order to make way for a newer facility by architect Valode & Pistre. It took advantage of the slopped terrain by building a concrete structure featuring a lower hall at the lower level and a large hall divisible into two at the upper level using a large mobile wall. Mechanical equipment called for 10 heat pumps for each hall subdivision; it was housed either at the roof level on the exhibition area side, or in a low recess under the ground floor level, also on this side. Noise control of the mechanical noise and activity noise radiated to the outside environment was of particular significance, as a powerful neighbor individual had managed to set a noise limit more stringent than the legal requirement.

The internal acoustics called for an absorptive treatment under the upper floor and on part of the wall. This made it possible to hold conferences with a sound system or large exhibitions. The ventilation was set at a reasonable 38 dB(A) so as to avoid being a nuisance in conference or banquet mode, while providing a bare minimum of masking noise in exhibition mode (with the possibility of generating a 45 dB(A) noise level by increasing the airflow rate).

15.7.4 Paris La Villette Salle Louis Armand

The science museum of Paris, known as the City of Science and Industry, is housed in a former modernistic slaughterhouse building in Paris. It features a conference facility by the designer Stark that is made of two 200-seat rooms separated by a 16 m wide mobile wall featuring a sound reduction index of 50 dB. The mobile elements of the floor allow for a 400-seat hall, or two 200-seat halls; according to the needs, the floor can be tiered or flat, with or without seats (e.g., for exhibition or banquet purposes).

Apart from the sound insulation between the two halls, the main challenge was the noise control of the mechanical equipment. The internal acoustics called for a partly reflective ceiling (so as to help propagate natural voices) and absorptive walls.

The facility enjoyed success to such an extent that the mobile wall was nearly daily operated. After 10 years of intensive use, the seals deteriorated to such an extent that leakage caused the sound reduction index to drop by 10 dB [4]. The refurbishment of the wall was carried out by its manufacturer Algaflex, and it returned to its former performances.

15.7.5 Le Vinci in Tours

The congress center of Tours is nicely located in front of the city railway station. Designed by architect Jean Nouvel, it features 2000-seat, 750-seat, and 350-seat halls, plus various meeting rooms.

While designed primarily as congress halls, those spaces can also be used for other purposes. The 2000-seat hall is often used for symphonic concerts, while the 350-seat hall is often used for chamber music.

The construction called for concrete floors and plasterboard partitions and ceilings. The acoustics of the larger venue can be adapted if need be using stage-like curtains on the walls.

Due to the large volume allocated to the mechanical equipment, HVAC noise control has been efficiently carried out.

15.7.6 Salle 3000 Amphitheatre in Lyon

Over the years the French city of Lyon built itself an "international city" featuring dwellings, offices, hotels, and a congress center. The architect of the whole city was the Renzo Piano Building Workshop. The masterpiece of it is the 3000-seat multipurpose hall erected at the end of the piece of land occupied by the international city and conveniently positioned as a signal by the Rhône River near the railway and the road arriving in Lyon from Paris.

This hall was designed as a congress hall; its shape was inspired by Roman amphitheatres (hence its name). As befits such a large facility, while its primary function is turned toward large plenary conferences and their opening or closing ceremonies, it can also house performances.

Sound insulation with regards to the outside environment is rather strong. To start with, the area can be a bit exposed to transportation noise with the busy rail lines out of the Lyon Part Dieu station. In addition, there are the highways on the bridge over the river and along the river too. But the environment must also be protected, as there are dwellings on the other side of the river. More to the point, the main façade of the building is turned toward the city and features a large double sliding door which separates the stage of the facility from the plaza in front of the hotels. That way, it is possible to hold a few performances, such as concerts, simultaneously with the performance hall and to the outside.

Due to the hemicylindrical shape of the hall, the acoustics of the place have called for a mixture of absorptive and diffusive treatments to prevent the occurrence of echoes.

Mechanical equipment has been housed partly on the roof and in plant rooms so as to comply with community noise requirements.

15.7.7 Salle Louis Frechette in Quebec

The 1952-seat Salle Louis Frechette in Quebec City, Canada, is intended for both music and theatre. It features a ceiling made of plate-like reflectors suspended from the ceiling, with an adjustable height for variable acoustics purposes. There also are motorized curtains along the walls, as well as motorized banners under the ceiling. There is a removable stage frame, as well as mobile proscenium walls, in order to adapt the stage opening to the hall in music mode.

An acoustic study of this hall was the subject of a paper [7]. When performing this study, the acousticians could verify that those dispositions were efficient, yet clearly enough, the proscenium walls were always kept in a position along the structural walls. When asked the reason, the stage staff happily answered that this was the best position because it left space on the side of the forestage to park the piano when it was not needed!

Lesson Learned: Always make sure the staff is aware of the purpose of the equipment!

15.7.8 Elgin Theatre in Toronto

The Elgin and Winter Theatres were designed by architect T. Lamb and opened in 1913. They are the last surviving Edwardian stacked theatres in the world. They were initially intended for vaudeville shows. After a small conversion of the lower theatre for sound films, the facility was eventually bought in 1981 by the Ontario Heritage Foundation and successfully housed *Cats* in the unrefurbished facility. Following a restoration from 1987 to 1989, a LARES system by Lexicon was installed at the request of the acousticians Muncy and Tanner. This system used two microphones on the balcony front edge to pick up sound from the stage, whose signals were digitalized and treated by two computers, with the resulting signals sent to 56 loudspeakers in the main ceiling and 60 under the balcony. This provided better intelligibility and ambience, with two main settings for either theatre or music.

15.7.9 Palais Omnisports de Paris Bercy

The Palais Omnisport de Bercy was initially built as a large indoor sport facility in Paris. However, due to its large size, it was quite tempting to hold performances in it (e.g., popular operas and pop events) as well as big rallies. Unfortunately, the acoustics of the facility had only been designed for sport events, and the audience and organizers were not overly thrilled.

A major renovation was eventually carried out. First, a diagnosis was performed [8] in order to assess the internal acoustics. Prescriptions followed suit, with emphasis on safety.

15.7.10 Palais des Congrès de Vichy

Vichy is a French town that used to be a fashionable thermal destination. As such, it features numerous hotels that make it perfect for congress organizers. A casino, featuring lounges and a small performance hall, had been built in the 1850s to entertain some of the rich guests, including Emperor Napoleon III. An opera hall was added later on (cf. Section 14.6.4). When eventually thermal stays went out of fashion, the casino operators found it more expedient to have a new, smaller facility built nearby. The township of Vichy bought back the premises and aimed to develop business travel to the city. This looked like a good plan, as it is only 3 h from Paris and less than 1 h from the industrial city Clermont-Ferrand.

However, the opera and grand casino were historical landmarks, so much caution had to be applied to the kind of work that could effectively be done. It was decided that the façades of the building would be kept in their original state, More to the point, the lounges that had paintings on the walls and ceilings had to be kept too. However, the small performance hall could be reworked, as it had been extensively modified since its inception. Last, the opera hall had to be kept as it was. While some palliative solutions could be developed for those spaces (in terms of both operations and fitting out), it was clear that modern spaces had to be created somewhere else. The break occurred when the architect in charge of historic buildings for the area reluctantly agreed that there used to be an entrance to the cellar on the park side. It was then deemed possible, through proper landscape planning on this side, to have an extra entrance and emergency exit, together with some natural lighting, thus enabling public use of the cellar to create modular meeting rooms, a restaurant area, and a communication room.

The meeting rooms were partitioned using 50 dB movable partitions and a moderately absorptive ceiling, so as to enable natural voices to be used. Fortunately, there was enough ceiling height to enable ventilation of the premises to be performed from the ceiling void. Adjacent partitions were absorptively treated in order to avoid flutter echoes, and most of the spaces in the former cellar were treated to an absorptive ceiling, with the noticeable exception of the communication room. In order to keep the beautifully vaulted upper floor, it was decided to use some carpet in the passageways but keep the stone flooring everywhere else, while specific absorptive panels were used to create cubicles along the walls and in the center of the room.

At ground level the heritage spaces were treated to curtains, with the restriction that a self-standing structure was needed for their suspension, as no drilling was allowed in the heritage walls. In the particular case of the opera hall, it was first envisioned to use infrared transmission to the audience using helmets. But as the facility was also meant to house such diverse activities as an opera, concert, and theatre, a thorough diagnosis was eventually performed [9] to help understand how a sound system could be implemented (still with the limitation that it had to be installed on a self-standing structure and no drilling was allowed).

REFERENCES

1. Peutz, Internal courses, 2014.
2. Peutz, De Maasport, Venlo (in Dutch), Report, 1986.
3. M. Barron, *Auditorium acoustics and acoustical design*, Taylor & Francis, Boca Raton, FL, 2009.
4. M. Asselineau, Acoustique active et acoustique passive—Approches complémentaires ou opposées? (Active and passive acoustics—Complementary or opposed ways? in French), presented at CFA2010 Proceedings, Lyon, 2010.
5. Peutz, Measurement report on a large mobile wall (in French), 1989.
6. M. Asselineau, The challenge of heavy rehabilitation projects—Case studies, presented at ICSV13 Proceedings, Vienna, 2006.
7. J.G. Migneron, M. Asselineau, Analyse acoustique du Grand Théâtre de Québec avec utilisation de l'intensimétrie acoustique pour l'évaluation des réflecteurs (Acoustical analysis of the Grand Theatre in Quebec City using acoustic intensimetry to assess the reflectors, in French), in *International Congress of Acoustics Symposium*, Vancouver, 1986, pp. 99–103.
8. A. Bradette, Practical and accurate room acoustical measurements in large indoor multipurpose halls and measures to optimize acoustics, presented at Proceedings of InterNoise 2013, Innsbruck, 2013, Paper 1067.
9. M. Serra, D. Chang, Diagnosis of an historical performance hall—Case study, presented at Proceedings of Acoustics 08, Paris, 2008.

Chapter 16

Conclusion

It is hoped that the reader had fun reading the various stories illustrating a few acoustic points of interest. I would like to emphasize the following points.

Before attempting to formulate a solution by means of prescriptions, one *must* formulate the problem. This means identifying the hopes of the end user, and the sensible or aggressive spaces in and around the project. Next, it is necessary to document the relevant regulations and standards.

To start with, any project starts with a program: What does the end user (and the payer too!) actually want? This means the various requirements must be identified and the relevant acoustic objectives stated. In today's spirit of sustainable development, one must be conscious that a team effort (i.e., architects, structural engineers, HVAC, and acoustics, just to name a few) must occur and make sure that the various solutions that are considered at the design stage are compatible with each other. Acoustics is often part of a global problem that can only be solved by a complete design team fed the relevant data by the end user.

Incidentally, standard practice (as defined in standards) may not always be the best practice (remember: By the time a consensus has been achieved for a standard to be published, new trends have probably emerged; more to the point, standards are the product of whoever cared to be present at the standardization meetings and may only represent part of a larger story). The design team is there to find out what the best course is.

A word of advice: Legal objectives are not necessarily synonymous with comfort (they are only meant to prevent too unhealthy situations from occurring); one should exercise one's own discretion according to the identified sensibilities of the end user.

Modeling is a nice tool, both for the specialist and for public information. Like all tools, it does have limitations and cannot be trusted blindly! While it is nice to establish a comparison between different situations—provided the model has been properly developed and tested—it seldom comes up with an absolute value to be trusted and cannot be operated by nonspecialists.

On completing the reading of this book, one should not expect to replace the acoustic specialist; however, one should be aware of the need to feed him with the relevant data.

Chapter 17

Example of a Building Project

17.1 DESCRIPTION

The following example features a type of building operation that is not uncommon in Europe, that is, a building housing dwellings and offices at the upper floors, and activities (workshop, shop, and restaurant) at the ground level. In addition, there are plant rooms at the ground level and roof level. On each side this building is bordered by dwellings. The layout of the operation is displayed in Figures 17.1 (ground floor), 17.2 (upper floors), and 17.3 (cross section).

We will first examine the implications regarding the acoustic targets of such a project, and then try to find the relevant acoustic prescriptions regarding sound insulation to the exterior, sound insulation between spaces, sound absorption, and mechanical equipment.

17.2 ACOUSTIC OBJECTIVES

17.2.1 Sound Insulation to the Exterior

To start with, the presence of a street in front of the building will most certainly require specific targets as per the legal requirements regarding the sound insulation of buildings exposed to road traffic noise (e.g., [1]) or the admissible background noise level inside buildings (e.g., [2]).

Next, one may consider the eventual noise likely to be generated inside the premises and radiated to the environment by the façades and roofs of the building: this means one will need to know the sound levels likely to be reached inside the premises (cf. Section 17.2.5). In addition, one will also need to know the background noise levels on site (cf. Section 17.2.6).

In this particular example, let's say that a sound insulation D_{nTAtr} of 35 dB is required for the exposition to road traffic noise.

17.2.2 Sound Insulation between Spaces

Minimal sound insulation values between spaces and maximal impact sound values between spaces are usually required by regulations (e.g., [3, 4]) for dwellings and sometimes for public receiving spaces too.

In this particular example, let's say that the following sound insulation D_{nTA} values and impact noise L'_{nTw} values are required:

Between dwellings: $D_{nTA} \geq 53$ dB and $L'_{nTw} \leq 55$ dB
Between dwellings and activity: $D_{nTA} \geq 58$ dB and $L'_{nTw} \leq 55$ dB
Between dwellings and mechanical: $D_{nTA} \geq 58$ dB and $L'_{nTw} \leq 55$ dB

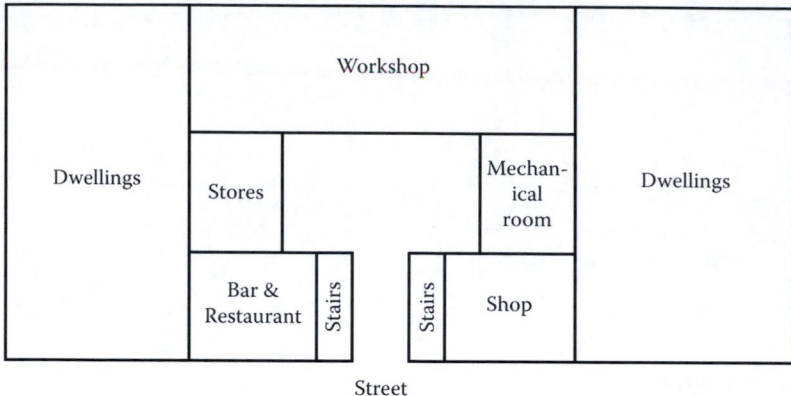

Figure 17.1 Schematic ground floor drawing.

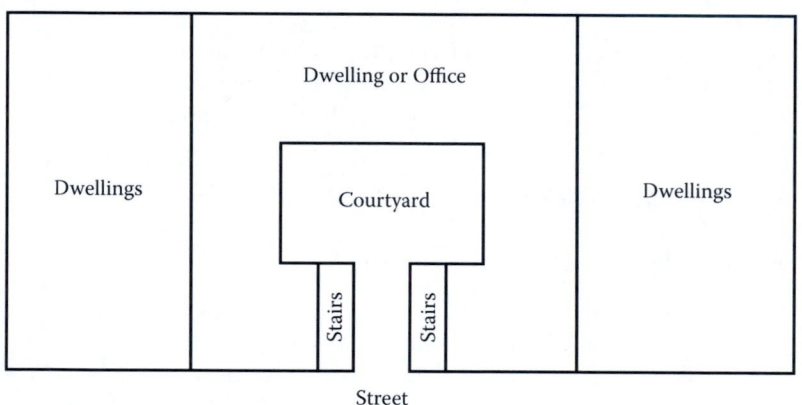

Figure 17.2 Schematic drawing of upper floors.

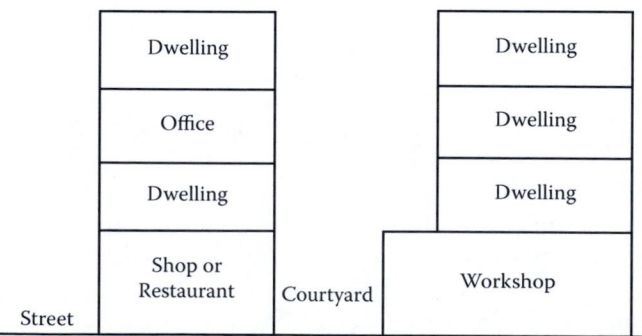

Figure 17.3 Schematic cross section.

In addition, some regulations may require specific sound insulation values between an activity space (e.g., community control regulations such as [5, 6]) and other places such as living quarters. More to the point, there may be extra requirements regarding the noise from leisure spaces (e.g., [7]). In both cases, it will be necessary to target a sound insulation value that is such that it ensures compliance with those regulations.

In this particular example, let's say that the emergence (gap between the ambient noise with the activity under operation) and the residual noise (background noise level) must not be greater than 3 dB in each octave band from 125 to 4000 Hz.

Within the offices, the user may want the possibility of partitioning his space according to needs. Let's say that the target between partitioned offices is:

With removable partitions: $D_{nTA} \geq 35$ dB and $L'_{nTw} \leq 55$ dB
With fixed partitions: $D_{nTA} \geq 45$ dB and $L'_{nTw} \leq 55$ dB

17.2.3 Sound Absorption

Some regulations require a minimum quantity of absorption inside common areas of dwellings (e.g., [3]). In addition, some regulations require a minimum quantity of absorption inside waiting areas of public receiving spaces (e.g., [8]). Last but not least, some occupational requirements may require a minimum quantity of absorption inside workshops (e.g., [9]).

In this particular example let's say that:

- An equivalent absorptive area representing 25% of the floor surface must be provided in the corridors and the entrance hall of the dwellings.
- An equivalent absorptive area representing 25% of the floor surface must be provided in the corridors and the entrance hall of the offices.
- A 3 dB(A) per doubling of distance spatial sound level decay must be provided in the workshop.

17.2.4 Mechanical Equipment

Mechanical equipment has to be considered from two angles: First, it may radiate noise to the environment, and second, it may also generate noise inside the premises.

According to its size and its nature (e.g., freezer, boiler, air handling unit (AHU)), mechanical equipment may be subjected to the noise community control regulations or to classified installations for the protection of the environment [10]. The relevant texts must be identified and their implications taken into account.

The noise levels generated by the mechanical equipment inside the premises are usually covered by regulations for dwellings [3], and often for public receiving spaces too [11]. Those must be listed and their implications taken into account. More to the point, some local regulations may require office spaces inside a predominantly dwelling-oriented building to have mechanical equipment whose acoustic characteristics comply with the regulations applicable to dwellings [12].

Incidentally, while the mechanical equipment chapter of the relevant regulations will consider the noise levels generated by such equipment, one must not forget about other implications, such as the sound insulation between spaces (that may be weakened by the duct network).

In this particular example, let's say that a background noise level of 30 dB(A) is required inside the dwellings.

17.2.5 User's Requirements

Occupational noise control legal requirements are of course applicable [13, 14].

Several activities are subjected to specific regulations; they must be listed and their implications must be indicated to the end user. In order to be able to properly specify the performance of the envelope of such spaces, it is necessary to assess the noise levels likely to be generated.

In some cases, it will probably be interesting (or even unavoidable!) to limit the sound level value permitted inside the premises. This must be pointed out to the end user, together with the relevant implications (e.g., operating at a higher sound level value will mean extra sound insulation precautions that will reduce the available floor space and increase the cost of the operation too).

One must also think of the future operation of the activities. To start with, a workshop or even a shop or restaurant will need deliveries; those will probably take place early in the morning, so a few specific precautions must be taken regarding the noise from delivery trolleys (e.g., avoiding large expansion joints in the floor covering to avoid bumps, requiring a trolley with resilient wheels, etc.). Next, there will be waste to be removed; this usually takes place late in the evening (with usually the same noise problems as deliveries).

Last but not least, some specific equipment will be needed by the operator (e.g., kitchen exhaust, freezers, production equipment, and sound systems). While the relevant user may have his own acoustician, the requirement book of the building operation must have specific requirements stated by the acoustician regarding noise limits (in terms of noise level values inside the operator's premises, sound insulation with regards to the other spaces, and sound levels radiated in the exterior environment or transmitted to the neighboring spaces).

In this particular example, let's say that a sound level of 80 dB(A) (with a pink noise spectrum) is permitted in the workshop, while a 85 dB(A) (with a musical noise spectrum) sound level is permitted in the bar-restaurant. In addition, a 3 dB(A) per doubling of distance spatial sound level decay must be provided in the workshop. More to the point, the ventilation equipment will not generate more than 60 dB(A) (this is not a regulation, but it will avoid being a strong contributor to the noise level exposure of workers).

17.2.6 Miscellaneous

In order to be able to specify the sound power level radiated by the mechanical equipment and by the façades (and in some countries, in order to be able to comply with the background noise level requirements inside the premises too), it is necessary to assess the background noise and the ambient noise levels on site.

There will probably be a standard describing the relevant measurement procedure (e.g., [15, 16]). Incidentally, one has to perform such a measurement—it is not sufficient to rely on a noise chart that will point out noisy areas in terms of daily L_{Aeq}, but will not provide indications regarding the background noise levels.

In this particular example let's say that a background noise level of 30 dB(A) has been measured inside the courtyard at nighttime, while the relevant value in daytime is 55 dB(A).

17.2.7 Construction

There usually are a few rules pertaining to the construction. In order to reduce annoyance to the neighborhood, there will typically be a specific time span to be complied with (e.g., 7:00 a.m. to 7:00 p.m. at most), with night work and weekend work allowed on a case-by-case basis. While the building permit will state the allowable hours of operation and the eventual restrictions (e.g., more stringent hours and sometimes the allowed methodology of demolition and construction), it may be necessary to apply a few restrictions according to the nature of the neighborhood (e.g., with a hotel next door, it might be appreciated to avoid starting noisy activities at first light!).

If a noise monitoring system has to be installed, the measurement points must be defined.

17.2.8 Summary

With due consideration to the various aspects reviewed in the previous chapters:

- Façade sound insulation:
  - $D_{nTAtr} \geq 35$ dB for the dwellings
  - $D_{nTAtr} \geq 35$ dB for the offices
  - $D_{nTAtr} \geq 30$ dB for the shops
  - $D_{nTAtr} \geq 45$ dB for the bar-restaurant
  - $D_{nTAtr} \geq 35$ dB for the workshop
- Sound insulation between spaces:
  - $D_{nTA} \geq 53$ dB between dwellings.
  - $D_{nTA} \geq 60$ dB between dwellings and activities such as offices, shops, and workshops. (Note: This is to take into account the noise from activities transmitted to the dwellings next door or above; the regular minimum value is only 58 dB.)
  - $D_{nTA} \geq 65$ dB between dwellings and bar-restaurant. (Note: This is to take into account the noise from activities transmitted to the dwellings next door or above; the regular minimum value is only 58 dB.)
- Equivalent absorptive area:
  - Corridors and the entrance hall of the dwellings: 25% of the floor surface
  - Corridors and the entrance hall of the offices: 25% of the floor surface
- Spatial sound level decay in the workshop of 3 dB(A) per doubling of distance
- Background noise level to be complied inside:
  - Dwellings: 30 dB(A). (Note: This is a legal requirement.)
  - Bar-Restaurant: 40 dB(A). (Note: This is a recommendation.)
  - Shop: 40 dB(A). (Note: This is a recommendation.)
  - Workshop: 50 dB(A). (Note: This is a recommendation.)
- Limitation of the noise levels, inside the facilities:
  - 80 dB(A) with a pink noise spectrum in the workshop, with operating hours in daytime only
  - 85 dB(A) with a music spectrum in the bar-restaurant, meaning a repartition of 92/89/85/83/79/77/71 dB in the octave bands from 63 to 4000 Hz, with possible operations in nighttime
- Background noise levels outside (based on measurements inside the courtyard):
  - 30 dB(A) in nighttime
  - 55 dB(A) in daytime

17.3 ACOUSTIC SPECIFICATIONS

17.3.1 Sound Insulation to the Exterior

17.3.1.1 Dwellings

The façade will be dimensioned on the shortest room, in our case, 2.50 m. Let's consider a 1.45 × 1.40 m² window. Under these conditions, the window will have to feature a sound

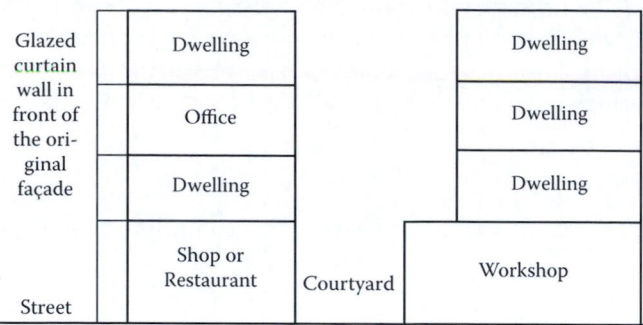

Figure 17.4 Schematic cross section of refurbished building with added-on glazed curtain façade wall.

reduction index $R_w + C_{tr}$ of 34 dB minimum inside a 15 cm thick reinforced concrete wall featuring a sound reduction index value $R_w + C_{tr}$ of 47 dB.

Should a light façade type be used, a global sound reduction index value $R_w + C_{tr}$ of 45 dB minimum will be needed.

Note: In addition to this requirement, the façade must enable the possibility of correctly separating different dwellings in compliance with the targeted legal sound insulation value. This will lead to a façade featuring a flanking sound insulation value $D_{ncw} + C$ of 60 dB minimum.

In addition, should the ventilation system be return air only, one should take into account the air inlets that will be located either in the façade main walls or in the window frame. The requirement for one such inlet will be D_{necw} of 38 dB.

Note: Under some regulations and bylaws it may not be permitted to use a standard façade when exposed to significant transportation traffic noise (e.g., [17]); it will then be necessary to introduce a curtain wall in front of the original façade (cf. Figure 17.4).

17.3.1.2 Offices

The required façade performance will be calculated for the office with the shortest dimension from the façade, in our case, 2.50 m.

Note: One should never calculate the required façade performance under the assumption that the floor of interest will be outfitted as an open-plan office with a significant depth behind the façade, as occupants close to it will not benefit from the same degree of protection as those in the middle of the floor.

This will lead to a façade featuring a sound reduction index value $R_w + C_{tr}$ of 38 dB.

Note: In addition to this requirement, the façade must enable the possibility of partitioning offices in compliance with the targeted sound insulation value. This will lead to a façade featuring a flanking sound insulation value $D_{ncw} + C$ of 45 dB minimum.

In addition, should the ventilation system be return air only, one should take into account the air inlets that will be located either in the façade main walls or in the window frame. The requirement for one such inlet will be $D_{ncw} + C$ of 35 dB.

17.3.1.3 Shops

The required façade sound insulation performance will be dimensioned according to the depth from the façade, in our case, 4.50 m.

This will lead to a façade featuring a sound reduction index value $R_w + C_{tr}$ of 30 dB.

In addition, should the ventilation system be return air only, one should take into account the air inlets that will be located either in the façade main walls or in the window frame. The requirement for one such inlet will be $D_{ncw} + C$ of 30 dB.

17.3.1.4 Bar-Restaurant

When solely dealing with the transportation noise requirements, the required façade sound reduction index can be calculated as we previously did with the office. However, we also want to manage the protection of the neighbors, with the closest located upstairs. Under the assumptions we made regarding the noise levels inside the facility, the required façade sound insulation of the bar-restaurant is 50 dB (Note: The neighbor is entitled to sleep with his window opened—hence the necessary degree of protection applied to the bar-restaurant façade).

The façade will be specified according to the depth from the façade, in our case, 5.00 m. This will lead to a façade featuring a sound reduction index value $R_w + C_{tr}$ of 45 dB.

Due to the complications involved, there will not be any air inlet in the façade.

17.3.1.5 Workshop

When solely dealing with the transportation noise requirements, the façade sound reduction index can be specified as we previously did with the office. However, we also want to manage the protection of the neighbors, with the closest located upstairs. Under the assumptions we made regarding the noise levels, the required façade sound insulation of the workshop then is 35 dB (Note: The neighbor is entitled to live with his window opened—hence the necessary degree of protection applied to the workshop).

The façade will be dimensioned according to the depth from the façade, in our case, 5.00 m. This will lead to a façade featuring a sound reduction index value $R_w + C_{tr}$ of 30 dB.

In addition, should the ventilation system be return air only, one should take into account the air inlets that will be located either in the façade main walls or in the window frame. The requirement for one such inlet will be $D_{ncw} + C$ of 41 dB.

One will also have to consider smoke exhaust trapdoors in the roof of the workshop. Those will have to feature a sound reduction index value $R_w + C$ sufficient enough (e.g., 45 dB).

17.3.2 Sound Insulation between Spaces

17.3.2.1 Dwellings

Walls between dwellings: 18 cm thick reinforced concrete wall featuring a sound reduction index $R_w + C$ of 57 dB.

Note: This would also have been possible using plasterboard partitions, but remember practical consequences (e.g., isn't it too tempting to drill a hole?).

Walls inside dwellings: 7 cm plasterboard partitions featuring a sound reduction index $R_w + C$ of 39 dB.

Floors between dwellings: 22 cm thick reinforced concrete slab featuring a sound reduction index $R_w + C$ of 61 dB and an impact noise L_{nw} of 69 dB. This floor will receive a floor covering featuring a sound reduction index ΔL_w of 13 dB minimum.

Floors between dwelling and office: 25 cm thick reinforced concrete slab featuring a sound reduction index $R_w + C$ of 63 dB and an impact noise L_{nw} of 59 dB. This floor will receive a floor covering featuring a sound reduction index ΔL_w of 13 dB minimum.

Floor between dwelling and shop: Ditto floor between dwelling and office.

Floors between dwelling and bar-restaurant: Under the assumptions made regarding the sound levels inside the premises, the use of a 30 cm thick reinforced concrete slab featuring a sound reduction index $R_w + C$ of 67 dB and an impact noise L_{nw} of 63 dB with a plasterboard ceiling underneath will not be sufficient to satisfy the requirements in the low-frequency range. For such a feat to be achieved, a box-in-box construction will be needed, with the metal structure supporting the plasterboard walls and ceiling attached to the floating slab.

Note: This will not only take space, but also require skilled workers.

17.3.2.2 Offices

Walls between offices: 15 cm thick reinforced concrete wall featuring a sound reduction index $R_w + C$ of 53 dB, or 16 cm thick plasterboard partitions featuring a sound reduction index $R_w + C$ of 59 dB.

Walls of the meeting room: 20 cm plasterboard partitions featuring a sound reduction index $R_w + C$ of 64 dB.

Door to the meeting room: Featuring a sound reduction index $R_w + C$ of 48 dB.

Walls to dwellings: 22 cm thick reinforced concrete wall featuring a sound reduction index $R_w + C$ of 61 dB.

Floors between offices: 15 cm thick reinforced concrete slab featuring a sound reduction index $R_w + C$ of 53 dB and an impact noise L_{nw} of 77 dB. This floor will receive a floor covering featuring a sound reduction index ΔL_w of 13 dB minimum.

Floors between dwelling and office: 25 cm thick reinforced concrete slab featuring a sound reduction index $R_w + C$ of 63 dB and an impact noise L_{nw} of 66 dB. This floor will receive a floor covering featuring a sound reduction index ΔL_w of 13 dB minimum.

17.3.2.3 Shops

Walls between shops: 15 cm thick reinforced concrete wall featuring a sound reduction index $R_w + C$ of 53 dB, or 16 cm thick plasterboard partitions featuring a sound reduction index $R_w + C$ of 59 dB.

Walls to dwellings: 22 cm thick reinforced concrete wall featuring a sound reduction index $R_w + C$ of 61 dB.

Floors of the shop: 25 cm thick reinforced concrete slab featuring a sound reduction index $R_w + C$ of 63 dB and an impact noise L_{nw} of 66 dB. This floor will receive a floating slab featuring a sound reduction index ΔL_w of 25 dB minimum.

Note: The floating slab is required in order to cope with the impacts generated by delivery trolleys. Should it be omitted, then no heavy deliveries will be possible during slumber hours.

Floors between dwelling and shop: 25 cm thick reinforced concrete slab featuring a sound reduction index $R_w + C$ of 63 dB and an impact noise L_{nw} of 66 dB. This floor will be treated to a ceiling featuring a sound reduction index ΔL_w of 10 dB minimum.

Note: This ceiling cannot be replaced by the absorptive ceiling of Section 17.3.3.3, as it has sound insulation and fire insulation capabilities.

17.3.2.4 Bar-Restaurant

Walls inside the facility: 18 cm thick reinforced concrete wall featuring a sound reduction index $R_w + C$ of 57 dB, or 20 cm thick plasterboard partitions featuring a sound reduction index $R_w + C$ of 64 dB.

Walls to dwellings: Under the assumptions made regarding the sound levels inside the premises, the use of a 30 cm thick reinforced concrete wall featuring a sound reduction index $R_w + C$ of 67 dB with two layers of plasterboard will not be sufficient to satisfy the requirements in the low-frequency range. For such a feat to be achieved, a box-in-box construction will be needed, with the metal structure supporting the plasterboard walls and ceiling attached to the floating slab.

Note: This will not only take space, but also require skilled workers.

Floors of the facility: 25 cm thick reinforced concrete slab featuring a sound reduction index $R_w + C$ of 63 dB and an impact noise L_{nw} of 66 dB. This floor will receive a floating slab featuring a sound reduction index ΔL_w of 25 dB minimum.

Note: The floating slab is required in order to cope with the impacts generated by delivery trolleys.

Floors between dwelling and facility: Under the assumptions made regarding the sound levels inside the premises, the use of a 30 cm thick reinforced concrete slab featuring a sound reduction index $R_w + C$ of 67 dB and an impact noise L_{nw} of 63 dB with a plasterboard ceiling underneath will not be sufficient to satisfy the requirements in the low-frequency range. For such a feat to be achieved, a box-in-box construction will be needed, with the metal structure supporting the plasterboard walls and ceiling attached to the floating slab (cf. Figure 17.5).

Note: This will not only take space and induce both structural and dimensional constraints, but also require skilled workers.

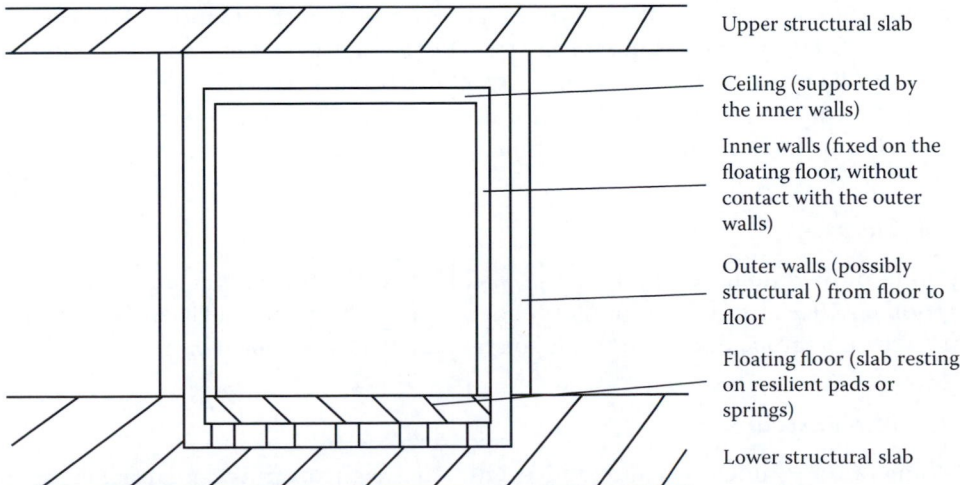

Upper structural slab

Ceiling (supported by the inner walls)

Inner walls (fixed on the floating floor, without contact with the outer walls)

Outer walls (possibly structural) from floor to floor

Floating floor (slab resting on resilient pads or springs)

Lower structural slab

Figure 17.5 Principle of box-in-box construction.

17.3.2.5 Workshop

Walls inside the facility: 15 cm thick reinforced concrete wall featuring a sound reduction index $R_w + C$ of 53 dB.

Note: This would also have been possible using plasterboard partitions, but remember practical consequences (e.g., isn't it too tempting to drill a hole?).

Walls to dwellings: 22 cm thick reinforced concrete wall featuring a sound reduction index $R_w + C$ of 61 dB.

Floors of the workshop: 25 cm thick reinforced concrete slab featuring a sound reduction index $R_w + C$ of 63 dB and an impact noise L_{nw} of 66 dB. This floor will receive a floating slab featuring a sound reduction index ΔL_w of 25 dB minimum.

Note: The floating slab is required in order to cope with the impacts generated by trolleys.

Floors between dwelling and workshop: 25 cm thick reinforced concrete slab featuring a sound reduction index $R_w + C$ of 63 dB and an impact noise L_{nw} of 66 dB. This floor will be treated to a ceiling featuring a sound reduction index ΔL_w of 10 dB minimum.

Note: This ceiling cannot be replaced by the absorptive ceiling of Section 17.3.3.5, as it has both sound insulation and fire insulation requirements.

17.3.3 Sound Absorption

17.3.3.1 Dwellings

The ceiling of the corridors and the hall will receive an absorptive material featuring an absorption coefficient of 0.70 minimum under 50% of the ceiling surface (note: this is a legal requirement). This can be achieved using acoustic tiles of such acoustic absorption.

17.3.3.2 Offices

The ceiling of the offices will receive an absorptive material featuring an absorption coefficient of 0.80 minimum under 100% of the ceiling surface. In addition, this ceiling will feature a flanking sound insulation D_{ncw} of 40 dB. This can be achieved using acoustic tiles of such acoustic performances (please note that the flanking insulation of the ceiling is supposed to be inclusive of the ventilation and lighting terminals embedded into it too).

17.3.3.3 Shops

The ceiling of the public accessible spaces will receive an absorptive material featuring an absorption coefficient of 0.70 minimum under 50% of the ceiling surface (Note: This may be left to the operator's discretion if it is so stipulated in the leasing contract).

17.3.3.4 Bar-Restaurant

The ceiling of the dining room and the bar will require an absorptive material featuring an absorption coefficient of 0.70 minimum under 50% of the ceiling surface (Note: This may be left to the operator's discretion if it is so stipulated in the leasing contract).

The ceiling of the kitchen and the dish washing room will require a cleanable absorptive material featuring an absorption coefficient of 0.70 minimum under 100% of the ceiling surface.

17.3.3.5 Workshop

The ceiling of the workshop will require an absorptive material featuring an absorption coefficient of 0.85 minimum using protected fiber tiles of such absorption under 100% of the ceiling surface (Note: This may be left to the operator's discretion if it is so stipulated in the leasing contract). In addition, one end wall above 1.0 m over the floor and the upper 25% of the main wall will require an absorptive material featuring an absorption coefficient of 0.85 minimum using fiber material protected by a glass cloth and a perforated metal panel of such absorption (Note: This may be left to the operator's discretion if it is so stipulated in the leasing contract).

17.3.4 Mechanical Equipment

17.3.4.1 Dwellings

If any, air intake inlets will feature a sound reduction index D_{new} of 41 dB.

The ducting system will feature primary silencers on air intake and exhaust, as well as supply and return air. While the exact dimensions will of course depend on the geometry of the layout and the sound power level of the AHU, an indicative length of 2 m is common.

Secondary silencers will be needed in order to prevent sound transmission between spaces by means of the ducting layout. Once again, while the exact dimensions will of course depend on the geometry of the layout, an indicative length of 2 m using an absorptive flexible duct and junctions that are at least 2 m distant from each other is common.

17.3.4.2 Offices

If any, air intake inlets will feature a sound reduction index D_{new} of 36 dB.

The ducting system will feature primary silencers on air intake and exhaust, as well as supply and return air. While the exact dimensions will of course depend on the geometry of the layout and the sound power level of the AHU, an indicative length of 2 m is common.

Secondary silencers will be needed in order to prevent sound transmission between spaces by means of the ducting layout. Once again, while the exact dimensions will of course depend on the geometry of the layout, an indicative length of 2 m using an absorptive flexible duct and junctions that are at least 2 m distant from each other is common.

In addition, there will probably be some smaller AHUs in subcompartments of the offices. The noise generated by these pieces of equipment that are usually located in the ceiling void depends, on the one hand, on the sound power level radiated by the equipment frame and shell, and on the other hand, on the sound power level generated by the air louvers, which are typically treated using 2 m of absorptive flexible duct (indicative length!).

In order to preserve the sound insulation properties of the ceiling, the air and lighting terminals will either be made of thick steel or be provided with an enclosure. Please note that a laboratory test report will be required from the contractor to prove compliance of the ceiling and terminal assemblies with the acoustic requirements.

17.3.4.3 Shops

If any, air intake inlets will feature a sound reduction index D_{new} of 36 dB.

The ducting system will be separate from the one of the dwellings. It will feature primary silencers on air intake and exhaust, as well as supply and return air. While the exact dimensions will of course depend on the geometry of the layout and the sound power level of the AHU, an indicative length of 2 m is common.

Secondary silencers will be needed in order to prevent sound transmission between spaces by means of the ducting layout. Once again, while the exact dimensions will of course depend on the geometry of the layout, an indicative length of 2 m using an absorptive flexible duct and junctions that are at least 2 m distant from each other is common.

17.3.4.4 Bar-Restaurant

The ducting system will be separate from the one of the dwellings. It will feature primary silencers on air intake and exhaust, as well as supply and return air. While the exact dimensions will of course depend on the geometry of the layout and the sound power level of the AHU, an indicative length of 2 m is common.

Secondary silencers will be needed in order to prevent sound transmission between spaces by means of the ducting layout. Once again, while the exact dimensions will of course depend on the geometry of the layout, an indicative length of 2 m using an absorptive flexible duct and junctions that are at least 2 m distant from each other is common.

A specific smoke exhaust layout will probably be required by the fire department. It usually cannot be outfitted with a silencer, and therefore it must not be used in normal operation. This means it will have a thoroughly insulated duct (for acoustic/thermal/fire purposes) and a slaved trapdoor at the bottom. In addition, there will be a need for fresh air; this can be achieved either through the entrance doors (but be aware, the fire department will often reject the use of air lock-mounted doors!) or through a specific set of slaved air supply doors that must be properly specified (e.g., here, a sound reduction index of 50 dB is a minimum).

17.3.4.5 Workshop

The ducting system will be separate from the one of the dwellings. It will feature primary silencers on air intake and exhaust, as well as supply and return air. While the exact dimensions will of course depend on the geometry of the layout and the sound power level of the AHU, an indicative length of 2 m is common.

A specific smoke exhaust layout will probably be required by the fire department. It usually cannot be outfitted with a silencer, and therefore it must not be used in normal operation. This means it will have a thoroughly insulated duct (for acoustic/thermal/fire purposes) and a slaved trapdoor at the bottom. In addition, there will be a need for fresh air; this can be achieved either through the entrance doors (but be aware, the fire department will often reject the use of air lock-mounted doors!) or through a specific set of slaved air supply doors that must be properly specified (e.g., here, a sound reduction index of 50 dB is a minimum).

17.3.5 User's Requirements

17.3.5.1 Dwellings

No such equipment provisioned for.

17.3.5.2 Offices

Office equipment will not generate more than 65 dB(A) at 1 m for heavy equipment (e.g., photocopier) and 55 dB(A) at 1 m for other equipment.

Note: This is to ensure there will not be too much noise inside the premises. Such a prescription is usually left to the tenant.

The usual requirement (in the contract) is for the tenant to comply with community noise control regulations (both for his activity and eventual equipment on the roof).

17.3.5.3 Shops

The usual requirement (in the contract) is for the tenant to comply with community noise control regulations (both for his activity and eventual equipment on the roof).

17.3.5.4 Bar-Restaurant

The minimal requirement (in the contract) is for the tenant to comply with community noise control regulations (both for his activity and eventual equipment on the roof). In addition, there will be a contractual requirement not to exceed the prescribed sound level values inside the authorized public spaces.

17.3.5.5 Workshop

The minimal requirement (in the contract) is for the tenant to comply with community noise control regulations (both for his activity and eventual equipment on the roof). In addition, there will be a contractual requirement not to exceed the prescribed sound level values inside the workshop.

17.4 CONCLUSION

This small example has given an idea of how complex the situation can be, as the acoustician will have to collaborate with (at least!) the architect, the structural engineer, the HVAC engineer, and the safety engineer, not to mention the client's and end users' representatives. Close cooperation between everybody involved is essential in such matters.

REFERENCES

1. Arrêté du 30 mai 1996 relatif aux modalités de classement des infrastructures de transports terrestres et à l'isolement acoustique des bâtiments d'habitation dans les secteurs affectés par le bruit (Arrest pertaining to the classification of ground transportation infrastructures and to the sound insulation of dwellings in areas affected by noise), *Journal Officiel de la République Française*, 149, 9694 (1996).
2. *Ontario noise assessment criteria in land use planning*, Publication LU-131, October 1997.
3. Arrêté du 30 juin 1999 relatif aux caractéristiques acoustiques des bâtiments d'habitation (Arrest pertaining to acoustic characteristics of dwellings), *Journal Officiel de la République Française*, 149, 10658–10660 (1999).
4. M. Morin, Qualification of degree of acoustic comfort in multi-family buildings, Report prepared by MJM Consultants for CMHC-SCHL, Montreal, Quebec, Canada, 1996.
5. Décret no 2006-1099 du 31 août 2006 relatif à la lutte contre les bruits de voisinage et modifiant le code de la santé publique (dispositions réglementaires) (Decree pertaining to community noise control), *Journal Officiel de la République Française*, September 1, 2006.
6. Calgary community standards bylaw SM2004, Part 9, Calgary, Alberta, Canada.

7. Décret n°98-1143 du 15 décembre 1998 relatif aux prescriptions applicables aux établissements ou locaux recevant du public et diffusant à titre habituel de la musique amplifiée, à l'exclusion des salles dont l'activité est réservée à l'enseignement de la musique et de la danse (Decree pertaining to prescriptions applicable to public receiving facilities or rooms usually featuring amplified music, with the exclusion of rooms for teaching music and dance), *Journal Officiel de la République Française*, December 16, 1998.

8. Arrêté du 1er août 2006 fixant les dispositions prises pour l'application des articles R. 111-19 à R. 111-19-3 et R. 111-19-6 du code de la construction et de l'habitation relatives à l'accessibilité aux personnes handicapées des établissements recevant du public et des installations ouvertes au public lors de leur construction ou de leur création (Arrest pertaining to handicapped people accessibility in public receiving buildings and facilities opened to public on their construction or creation), *Journal Officiel de la République Française*, 195 (2006).

9. Arrêté du 30 août 1990 pris pour l'application de l'article R. 235-11 du code du travail et relatif à la correction acoustique des locaux de travail (Arrest pertaining to the acoustic treatment of workplaces), *Journal Officiel de la République Française*, 224 (1990).

10. Arrêté du 23 janvier 1997 relatif à la limitation des bruits émis dans l'environnement par les installations classées pour la protection de l'environnement (Arrest pertaining to classified installations for environment protection), *Journal Officiel de la République Française*, March 27, 1997.

11. Arrêté du 23 juin 1978 relatif aux installations fixes destinées au chauffage et à l'alimentation en eau chaude sanitaire des bâtiments d'habitation, de bureaux ou recevant du public (ERP) (Arrest pertaining to fixed installations for heating and hot water providing of dwellings, offices, or public receiving), *Journal Officiel de la République Française*, 5606 (1978).

12. Circulaire du 9 août 1978 relative à la révision du règlement sanitaire départemental type (Circular dated August 9, 1978, pertaining to the revision of the typical district sanitary regulation), *Journal Officiel de la République Française*, 7188 (1978).

13. Décret n° 2006-892 du 19 juillet 2006 relatif aux prescriptions de sécurité et de santé applicables en cas d'exposition des travailleurs aux risques dus au bruit et modifiant le code du travail (Decree about safety and health prescriptions applicable in case of workers exposed to noise induced risks and modifying the work code), *Journal Officiel de la République Française*, 166, 10905 (2006).

14. Alberta Occupational Health and Safety Code, Section 218.

15. ISO 1996-1: *Acoustics—Description measurement and assessment of environmental noise—Part 1: Basic quantities and assessment procedures*, Geneva, 2003.

16. AFNOR NF S31010: *Acoustique—Caractérisation et mesurage des bruits de l'environnement—Méthodes particulières de mesurage* (*Acoustics—Characterization and measurement of environmental noise—Peculiar methods of measurement*), St. Denis, 1996.

17. C. Schoonebeek et al., Urban planning and quiet places in Amsterdam, QSIDE (LIFE09 ENV/NL/000423), http://www.qside.eu/nweb/urban_amsterdam.html.

Chapter 18

Examples of Fitting Out

18.1 DESCRIPTION

Here are simple examples of fitting out inside an existing building. Compared to the example of Chapter 17 involving the construction or heavy rehabilitation of a complete building, we are dealing with the construction or rehabilitation of a room inside an existing building. Two simple examples are given: a bedroom and a meeting room.

We will first examine the implications regarding the acoustic targets of such a project, and then try to find the relevant acoustic prescriptions regarding sound insulation to the exterior, sound insulation between spaces, sound absorption, and mechanical equipment.

18.2 BEDROOM

18.2.1 Description of the Operation

The owner of a flat has decided to improve the sound insulation of his bedroom. The initial brief simply says (polite version): "We do not want to hear the folks from the upper floor!" The location of the tenant's bedroom with regard to other spaces is displayed in Figure 18.1.

18.2.2 Diagnosis and Feasibility

First, an acoustic diagnosis is performed. It aims to ascertain by means of acoustic measurements the background noise level value inside the bedroom of interest, as well as the sound insulation value between this bedroom and the spaces around it (including the outside environment). It also aims to point out the potential weaknesses that can be observed in the construction (e.g., ventilation duct linking two spaces, poor quality separating wall or ceiling, etc.). One should note that the advice from a structural engineer will probably be needed too in order to ensure that the extra load is compatible with the existing structure.

On the basis of this diagnosis, the acoustician will submit a set of acoustic objectives covering the sound insulation to the outside D_{nTAtr}, the sound insulation to other spaces D_{nTA}, the reverberation time, and the noise from mechanical equipment. Most of the time, in existing buildings, a gain of 5 dB(A) minimum over the existing situation will be sought. If not practical, the end user will be informed.

18.2.3 Prescriptions

18.2.3.1 Sound Insulation

First, holes (e.g., from previous mechanical equipment or fixations) will be sealed.

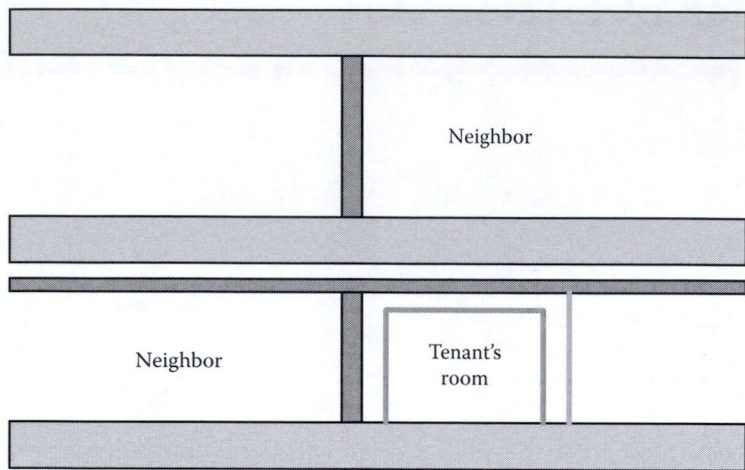

Figure 18.1 Schematic cross section of the tenant's room and neighbors.

The most efficient acoustic way of improving the sound insulation of the bedroom is to prevent noise radiation from its envelope.

There is not much hope of achieving anything on the floor, as it would mean a floating floor, with a step to enter the bedroom. However, should such an improvement be sought, there are manufacturers providing heavy wooden panels assembled on resilient pads and featuring a mineral wool glued on their underface (e.g., [1]). Please kindly note that a 10 cm height is needed for such a construction.

Note: It usually is ill advised to try to pour concrete on the existing floor. To start with, it will considerably increase the weight per floor area (to an extent not necessarily compatible with structural limitations), and more to the point, it will result in a stiffer floor that may generate a higher impact noise.

Next, a metal structure will be built in order to support the additional walls and the ceiling. There are manufacturers providing strong enough structural elements to avoid a structural connection with the upper floor and the walls (e.g., [2]). If a floating floor has been chosen, then those structural elements will be mounted on it rather than on the structural floor. Plasterboard plates will be screwed on this metal structure, with a 100 mm mineral wool between structural elements.

Note: Already, 35 cm has been eaten up on both horizontal dimensions of the room and the vertical dimension.

A new window will be built and attached to the inside plasterboard walls. It will have to be wider than the regular window in order to allow for its opening. It will typically feature a sound reduction index $R_w + C$ of 30 dB.

Note: The actual required performance will depend on the improvement targeted on the sound insulation to the other spaces. Remember, there is a need for balance between the various noise contributions!

Note: Using a floating floor, it also means that the door-window, if any, will have to be replaced due to the step to be introduced.

A new door to the corridor will be built and attached to the inside plasterboard walls. It will have to be wider than the regular door in order to allow for its opening. It will typically feature a sound reduction index $R_w + C$ of 30 dB.

Note: The actual required performance will depend on the improvement targeted on the sound insulation to the other spaces. Remember, there is a need for balance between the various noise contributions!

Note: Using a floating floor, it also means that the original door will have to be replaced due to the step to be introduced, that is, unless this door is made to open to the corridor, but do be careful on opening due to the risk of bumping into it when going into the corridor!

18.2.3.2 Sound Absorption

Usually in a bedroom, additional sound absorption is not sought due to the amount already provided by the furniture.

18.2.3.3 Mechanical Equipment

Silencers will be installed on the supply and return ducts in order to comply with both the sound insulation target and the background noise target. Care will be applied at the junction of the ducts on entering the room through the partition to preserve the sound reduction performances of the partition.

Supply and return louvers will be specified in order to limit their sound power level to a value compatible with the background noise level target (as an indicative value under the assumption that there will be one return and one supply louver, a sound power level of 35 dB(A) is advisable when taking into account the distance).

Should an air inlet be provided in the existing façade, another inlet (typically featuring a D_{new} of 35 dB minimum) will have to be provided in the new window assembly.

18.2.3.4 User's Equipment

The usual end user's previous equipment should be left (e.g., clock), as it participates in the balance between noise contributions inside the room.

18.2.4 A Few Comments

It is not uncommon in old buildings to find that the improvement that was dreamed of by the end user cannot be met. There can be a variety of reasons for that; just to name a few: The structure of the building can be too weak to bear the extra load needed by the rehabilitation, the appearance of the façade cannot be altered with impunity due to the architectural classification of the building, the space eaten up by the new construction is too much, etc.

By the way, please kindly remember that in addition to the usual legal requirements regarding rehabilitation work, there will probably be a condominium rule applying too, with, for example, restrictions on the working hours and specifications on the nature of the work permitted.

Please kindly remember that rooms are supposed to be ventilated (there are usually legal requirements on that topic). Therefore, supply air (through façade inlets or separate ducting) and return air (through separate ducting) must be provided.

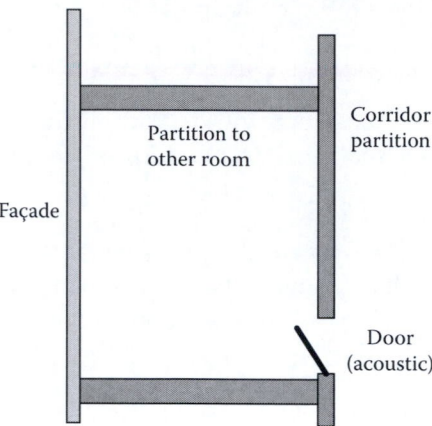

Figure 18.2 Layout of the meeting room and nearby spaces.

18.3 MEETING ROOM

18.3.1 Description of the Operation

The management has decided to have a new meeting room on a floor of an office building. The initial brief simply says: "We do not want to hear anything!" The location of the meeting room with regard to other spaces of the office floor is displayed in Figure 18.2.

18.3.2 Diagnosis and Feasibility

First, a meeting is held with the client's officer responsible for construction work inside the building. One has to explain to the management that "not hearing anything" is a bit complicated and awfully expensive. After pointing out the standards' recommendations, the acoustician gives a choice of privacy level (meaning a sound insulation of 45 dB at a minimum) or security level (meaning a sound insulation of 55 dB at a minimum). He also points out right from the start that removable partitions are out of the question due to their poor acoustic performance. Eventually, the client decides on a privacy level quality.

The acoustician defines a background noise level inside the meeting room (35 dB(A) from mechanical equipment) and a reverberation time (1.0 s in the 1000 Hz octave band).

Prior to the operation taking place, the acoustician checks through a diagnosis on the existing rooms whether the acoustic performances of the façade and floors are compatible with those objectives (e.g., if the façade or the floors are too weak, there will be such a flanking transmission that the sound insulation target will not be met). He will also look for such singularities as a thermal insulation along the façade (in the affirmative, he will have to further check whether it is made of mineral wool or only of simple polystyrene) or a technical space along the façade (that will have to be closed later on). More to the point, with the help from maintenance people or his colleague from HVAC, he checks that there is no ducting network transiting through the ceiling and floor voids.

Once this has been checked, a meeting is held between the acoustician and the end user. If some improvements are needed on the existing floors and walls, they are then identified. A last check is made to ensure that everything wanted, from the number of seats and tables to the video equipment and the eventual specific air handling unit, can be accommodated in compliance with the regulations (including safety requirements).

18.3.3 Prescriptions

18.3.3.1 Sound Insulation

The walls of the meeting room will be made of a 20 cm thick plasterboard partition mounted from floor to floor and featuring a sound reduction index R_w of 64 dB. This means that the ceiling and the technical floor will have to be interrupted on each side of the partition. Care will be applied at the junction of the partition with the façade.

Note: Mounting the partition from a technical floor to the ceiling will normally not allow a privacy level performance due to flanking transmission through these constructive elements.

Note: If there is a thermal insulation complex along the façade, it will most probably have to be interrupted at the partition location to allow compliance with the acoustic targets, but this will have to be checked with the HVAC engineer (to preserve the thermal insulation of the premises) and the safety engineer (for fire containment purposes).

The door to the meeting room will feature a sound reduction index R_w of 50 dB.

Should the upper-floor performance be insufficient (that should have been identified during the diagnosis!), a plasterboard and mineral wool ceiling assembly of the required performance will be suspended from the upper floor.

18.3.3.2 Sound Absorption

The ceiling will be absorptive. It can be made of absorptive tiles featuring an absorption coefficient α_w of 0.80 minimum.

Note: If a plasterboard ceiling has previously been mounted under the upper floor, the absorptive tiles will have to be either glued or fixed close to this ceiling (there are special hangers for that purpose with some manufacturers, e.g., [3]).

In addition, the wall on the corridor side and the side wall opposite of the projection screen will feature an absorptive treatment with an absorption coefficient α_w of 0.60 minimum. This can be achieved using perforated wood or metal panels (with a 18% minimum perforation ratio) over a 5 cm minimum mineral wool.

In order to enhance voice propagation inside the room, a reflective panel will be hung over the table. This panel will feature a convex shape in order to avoid a flutter echo between the table and panel. The location of those treatments is displayed in Figure 18.3.

The floor will be treated to a carpet featuring an absorption coefficient α_w of 0.15 minimum.

18.3.3.3 Mechanical Equipment

Silencers will be installed on the supply and return ducts in order to comply with both the sound insulation target and the background noise target. Care will be applied at the junction of the ducts on entering the room through the partition to preserve the sound reduction performances of the partition.

Supply and return louvers will be specified in order to limit their sound power level to a value compatible with the background noise level target (as an indicative value under the assumption that there will be one return and two supply louvers, a sound power level of 40 dB(A) is advisable when taking into account the distance).

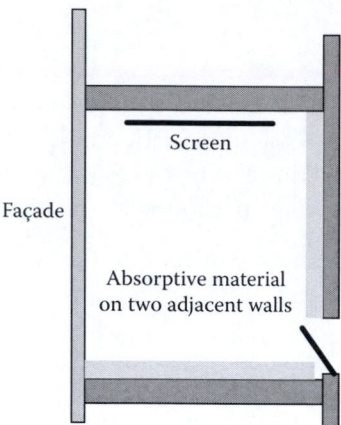

Figure 18.3 Location of acoustic treatments in the meeting room.

18.3.3.4 User's Equipment

Care will be exercised when choosing equipment such as the video projector and the computers to be installed in the room. A sound power level of 50 dB(A) maximum is advisable.

18.3.4 A Few Comments

The best removable partitions, when tested in laboratory conditions, feature a sound reduction index R_w of 45 dB. However, it should be remembered that under the actual site conditions, the floors are not truly horizontal, the façade may not be fully vertical, and the connecting constructive elements are not as massive. In practice, due to the leakage involved, it is not realistic to pass the 38 dB sound insulation mark with such elements.

Kindly remember that a 50 dB door is heavy and can be hard to operate. When space is available, it is much better to have an air lock that will allow simpler doors.

Absorptive treatment on two adjacent walls is meant to eliminate nasty flutter echoes inside the room. While the reverberation time target could be met without such a wall treatment, it does enhance speech intelligibility inside the room.

Achieving good speech intelligibility in the room also implies that there is not too much background noise. This means that the HVAC must be controlled.

More to the point, the end user must be warned regarding the type of equipment he is bringing inside the meeting room, as the average low-grade video projector will generate too much noise for comfort when unchecked.

REFERENCES

1. CDM, Isofloat floor system, Commercial brochure, 2013.
2. Placo, Megastil high performance plasterboard partitions, Commercial brochure, 2013.
3. Ecophon, Z hangers, Commercial brochure, 2013.

Chapter 19

A Word about Other Interesting Topics (Thermal and Fire Protection of Buildings, Structures, and Rodent Repulsion)

19.1 INTRODUCTION

The aim of this chapter is just to remind the acoustically minded that there are other fields strongly interacting with acoustic design. It is wise to check their requirements in order to avoid painful misunderstandings resulting in lost time and energy.

With acoustics being a so-called transversal discipline, there will be interactions with most of the building fields, for example, as innocuous as painting, as paint can eventually ruin an absorptive material.

19.2 STRUCTURAL ENGINEERING

It is well known to the acoustician (and most of the time to the architect too) that a heavy wall or floor is often a simple and useful way to obtain a rather high insulation wall or floor [1]. But can it always be obtained?

The structural engineer wants the building to support itself. He also wants it to be able to comply with some specific requirements. For example, the engineer may be confronted with a situation where there is high seismicity; this will result in the need to lighten the top of the building (the acoustician will probably end up with a lighter than hoped for floor slab) and a larger than planned expansion joint (typically 10 cm wide, which can be a problem for both acoustic insulation and fire control). More to the point, he may want the structure to support itself under special circumstances (e.g., earthquake, fire), and this will have some consequences for the construction too, especially if vibration control measures have to be implemented (cf. Section 19.6.1).

Regarding the weight aspects, this usually is not too much of a problem if identified early on; one will then use plasterboard ceilings and partitions (cf. Sections 12.5.4 and 3.12.14) to achieve the required sound insulation between floors. But one will have to be careful regarding impact noise and vibration control: Should the floor be insufficiently stiff, then the vibration attenuation measures will not work properly. This may eventually result in beams under the equipment to be resiliently supported, but please kindly note that in real life, it is not uncommon to shift the location of the technical and mechanical equipment during the project (with awkward problems then to be solved supporting the equipment!). More to the point, the presence of large beams may be a problem with regards to the available ceiling height.

19.3 THERMAL INSULATION

19.3.1 Scope

How come thermal insulation might be a problem to the acoustician? Well, there are quite a few points to beware of: To start with, quite a number of people will confuse an acoustic absorptive material and a thermal insulation layer, resulting in both materials being familiarly called insulation, with sometimes disastrous consequences (cf. Section 3.12.11).

Managing thermal comfort has been a constant requirement for man. While efforts were initially directed at heating systems, the scarcity of energy sources has prompted the legislator to issue regulations regarding the thermal insulation of buildings (e.g., [2–4]).

19.3.2 Thermal Insulation

Thermal insulation is the prevention of the passage of heat [5]. It may have to deal with the three modes of heat transfer:

- Conduction
- Convection
- Radiation

Usually a reflective or low-emittance sheet will not be considered a thermal insulation material, though it will reflect heat. Likewise, a radiant barrier or a vapor retarder will not be considered a thermal insulation material.

The thermal resistance R of a material is expressed in m²K/W. The higher the value of R, the better an insulation material it is. Its thermal conductivity K (λ in some countries) is expressed in W/m.K and characterizes the quantity of heat that can be transferred into a material within a given time. The smaller the value of K, the more insulating the material is; insulating materials are considered to feature values under 0.06 W/mK.

The heat transfer coefficient U characterizes the quantity of heat through a wall under established flow per time unit, surface unit, and temperature unit differences between the ambiances on each side of that wall. It is expressed in W/m²K and is the inverse of the total thermal resistance of the wall. The lower its value, the better the insulation is.

Insulating the wall seems straightforward enough, doesn't it? One simply has to apply a thermal insulation layer; however, applying that layer may result in downgrading the sound reduction index of the wall. For example, applying plasterboard and polystyrene doubling on a brick wall will usually result in a 4 to 7 dB loss on the sound reduction index. More to the point, applying a doubling on the inner surface of the façade will complicate matters: The HVAC engineer will look forward to having a well-insulated envelope and will not accept that his interior thermal doubling will be interrupted. The acoustician will usually hate this particular doubling, as it will induce flanking transmissions; he will look forward to pushing the partition to the façade (and incidentally, so will the safety engineer for fire control purposes), but this will create a thermal bridge (cf. Section 19.6.3). Another sore point will be the connection of a curtain façade to the floors: The HVAC engineer will love the use of thermal bridge breakers, but both the acoustician and the safety engineer will loathe it, as it will create a secondary transmission path between floors (you have probably guessed by now that it will require mineral wool and a plate top and bottom, as displayed in Figure 19.1, to ensure the desired acoustic and fire result).

Incidentally, applying an acoustically absorptive material on a wall where there already is thermal insulation may seem a waste of space and money to the architect or the end user.

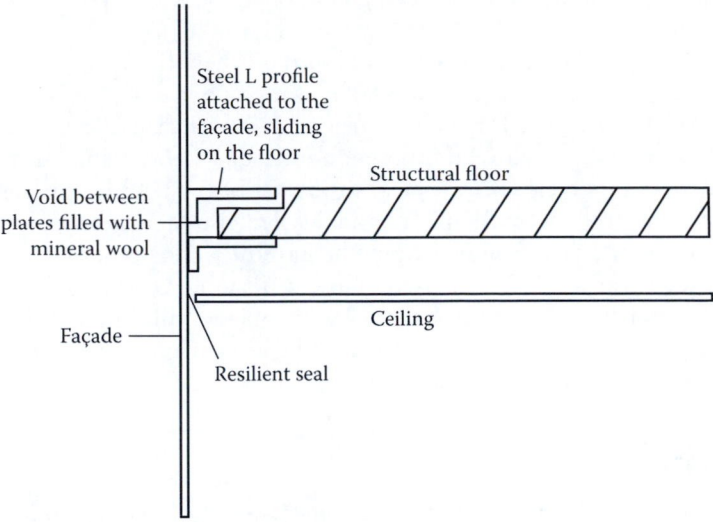

Figure 19.1 **Example of floor and façade assembly.**

Often a decision is made to implement a single material instead of the usual double layer of insulation material (two-thirds of the total thickness) and acoustic absorptive material (one-third of the total thickness, as displayed in Figure 19.2).

Unfortunately, this results in a material with a vapor barrier on the room side that will ruin the absorptive characteristics of the wall covering.

It is quite fashionable nowadays to make use of the thermal inertia of a building. This means that the walls and ceiling must be kept bare, but how does one achieve the required acoustic absorption then? Well, one could think of an acoustically absorptive material with a high thermal conductivity (e.g., a metal foam), but there are problems associated with its fixation on the walls and ceiling; more to the point, it may complicate matters when dealing with fire safety too. While such materials have actually been developed for specific applications (e.g., the absorptive interior lining of an electronic cabinet where one must try to

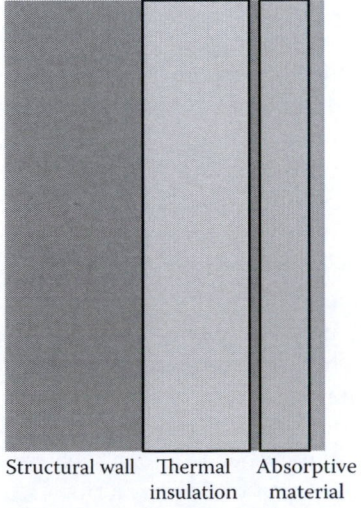

Figure 19.2 **Example of the principle of thermal insulation combined with acoustic absorption.**

reduce the noise while simultaneously taking care of thermal dissipation through its metal walls), their cost is yet prohibitive for regular building use [6, 7].

Another way is to use the convection effects to the fullest by having a suspended ceiling made of horizontal panels that covers only 60% of the upper floor surface [8].

Baffles can also be considered for such a purpose. However, there usually is a bit of a fight between the HVAC engineer, who will argue for a rather wide spacing, and the acoustician, who will prefer them much closer (a typical ratio will be 1 m² of baffle per 1 m² of floor). More to the point, the architect will understandably prefer to achieve a decent ceiling height, so the resulting expected performances must be carefully considered (cf. Section 19.6.13). On the other hand, baffles can also be used on the walls of large spaces (cf. Section 19.6.14).

Incidentally, there may be noise associated with objects dilating or contracting. While those objects will often be ducts or some façade elements, some larger objects may also be concerned. Section 19.6.5 illustrates the point.

19.3.3 Ventilation

The HVAC engineer is highly motivated to implement ventilation of the building spaces, as it will prevent humidity and moisture. Unfortunately, the simplest way is to locate air inlets in the façade (one usually will find such appliances in the window frames or the roller shutter box if there is not too much sound attenuation required). Such an inlet eventually turns out to be a mess for both the HVAC and the acoustic engineer: Heat is lost that way, and noise enters there too. The fact is, while elements are counted in calculations using their surface and sound reduction index or thermal resistance, leakage quickly becomes one of the major weaknesses for both the HVAC engineer and the acoustician. Gone are the days when in order to ventilate a building, one would simply open the window and let both noise and fresh air in! When there is no such high noise exposure as to require full mechanical ventilation, one usually uses air inlets of suitable attenuation; the less efficient will simply feature a small absorptive hood on each side of a slit in the window frame, while the more efficient will usually feature a small silencer and must be accommodated in the wall. On the other hand, there have been developments attempting to use regular windows behind a glazed ventilated enclosure or even a glazed screen [9, 10]. Some projects are actually calling for a double façade [11], but one must be careful regarding both acoustics (due to the high risk of transmission between spaces via the façade void) and fire prevention (due to the potential chimney effect).

Actually, both the acoustician and the HVAC engineer share a common concern regarding possible weaknesses in the envelope of the building (e.g., windows, air inlets, etc.). This means that when performing a diagnosis of an existing building, they can supplement each other to find where the effort must be [12]. However, problems may arise later as both engineers are working on the same building elements with slightly different motivations; however, with careful coordination, it is perfectly possible to achieve the required result in both fields.

In quiet areas, natural ventilation should please the acoustician, as there will be no noise contribution from the ventilation system! Unfortunately, ventilation noise may be required by the acoustician as masking noise in such places as open-plan offices or waiting areas (cf. Section 19.6.3). Should such a scheme be contemplated, the HVAC engineer and the acoustician had better coordinate with each other, as the background noise levels may turn out to be insufficient to meet the acoustic requirements regarding the minimum background noise level value. This may either rule out such concepts as thermoactive slabs or require specific masking noise precautions. In addition, when looking simultaneously for compliance with several sustainable development certifications (e.g., HQE (Haute Qualité Environnementale, French label for high-quality sustainability) and BREEAM (Building

Research Establishment Environmental Assessment Methodology)), and LEED (Leadership in Energy and Environmental Design) stringent requirements in several fields, and the margin allowed can be quite small indeed (e.g., for an office space to be fitted out, under normal operating conditions HQE will require a maximum background noise level of 43 dB(A), while BREEAM will require it to be in the 40 to 50 dB(A) range).

19.3.4 Sunscreens

Sunscreens may seem innocuous enough as to be ignored by the acoustician. However, they often can turn out to be a nightmarish affair.

Let's start with the small indoor screens: When manually actuated, their noise may be rather low. But unless properly designed and installed, when they are electrically operated this may be a bit of a problem should the neighboring rooms be looking for some quietness! It is advisable to have a prototype mounted and tested prior to committing to a full order for the whole building.

Note: Beware that most of the time any mechanical defect will result in higher noise levels.

Outdoor screens are a quite different matter: Due to their exposure to the elements, they have to be of sturdier construction. When they are mobile, the mechanisms are more powerful than those for indoor use, and they may generate vibrations in the structure and ultimately noise in the rooms nearby. More to the point, the wind may generate turbulences on such singularities, which can generate noise and vibrations under specific speed and incidence conditions (cf. Section 19.6.4).

Note: While there are some basic formulas to predict the sound power level of a grid under airflow pressure, nothing will replace an acoustic test. It should be noted that usually the façade manufacturer will have a wind tunnel test scheduled in his prescription book; unfortunately, most of the time such wind tunnels are used to find, for structural purposes and a given incidence and speed, what the resonance frequency value is and what the corresponding amplitude is. The noise generated by the screen will usually be drowned in the noise of the fan used to simulate the wind. It is compulsory to use an acoustic wind tunnel facility, which features silencers to reduce fan noise.

19.3.5 Diagnosis

The typical purpose of a diagnosis is to find how to save on fuel consumption, as this is a financial incentive. In order to better sell their heating products (ranging from fuel burners to electric heaters), a few service companies have quickly found that it was commercially interesting to offer a potential customer a diagnosis package, including both thermal aspects (i.e., the thermal insulation of the building envelope and the means of ventilating and heating or cooling the premises) and acoustic aspects.

There definitely are similarities between both aspects: A leakage will be both a way in for noise and a way out for the heat. However, there also are a few differences: An acoustician can be happy with a thick single glass pane, while the HVAC engineer will go berserk!

Improvements as envisioned during the diagnosis will also differ: The HVAC engineer may be happy with a new shutter system including an air intake, but the acoustician will usually balk at it due to the light construction and lack of attenuation with regards to the outside. Similarly, the HVAC engineer will usually prescribe a new thermal insulation on the outer side of the façade, which means that on a light façade, acoustic flanking transmissions will be a problem.

Performed by a person knowledgeable in both thermal and acoustic aspects, a diagnosis, even simplified, can point to potential problems and pave the way for the prescriptions aimed at an efficient rehabilitation (or simply an improvement of the thermal and acoustic comfort) [12].

19.4 FIRE PROTECTION

This section does not intend to explain all the fine points of fire protection. It simply aims to point out a few items of interest.

19.4.1 Design

19.4.1.1 Fire Detection

Legal requirements vary from one location to another. The first thing is to take stock of those requirements. It must be emphasized that there are specialists known as safety engineers who are qualified in this field. They will help point out the issues and propose solutions to avoid problems. More to the point, those specialists usually know the firemen well and will be able to analyze their needs and make sure the project does answer every one of them.

In public receiving buildings, and more and more often in dwellings too, there is a fire detection system. Fire detection can be performed using three techniques:

- Smoke detection (opacimetry)
- Heat elevation
- Color of flame

According to the sensibility of the space, detection may be validated on two or even three criteria being simultaneously verified (low-sensitivity case) or on any one of them (high-sensitivity case).

Those detectors must be kept free from any obstacle nearby and fully exposed to the view from any point of the zone they are supposed to cover; this means that they cannot be hidden between two absorptive baffles.

In addition, the fire alarm can usually be triggered through a manual switch available in the corridors and hallways. The responsible personnel (often the receptionist at the desk) must be trained to identify an alarm and rearm the system in case of wrongful alarm. In most public receiving buildings the firemen will automatically be notified after a certain delay. More to the point, a hotel or cinema operator is usually not keen to find his clients thrown in the street because of a faulty alarm, so training of the relevant personnel is a must.

19.4.1.2 Fire Alarm

In order to warn the occupants of the building, an alarm usually is sounded (usually supplemented with a luminous signal), and for public receiving facilities, a voice message is heard over the PA system. According to the regulations in force (as well as the local habits of the safety personnel in charge!), there are requirements on the PA system used for that purpose. In Europe, a common intelligibility scale (CIS) of 0.7 minimum is required by the European standard on PA systems for emergencies [13], which is quoted in the relevant European regulations. The same value is also requested by the National Fire Protection Association (NFPA) in the United States. In addition, through lack of measurement capability or for simplification purposes, the safety personnel often add an audibility clause (with the alarm signal being required to be 10 or even 15 dB over the background noise), so that the alarm signal can be heard whatever the circumstances [14]; that is not without some consequences on the design (cf. Section 19.6.12).

Note: Please kindly remember that while the alarm must be audible everywhere, it must not blast the eardrums of people nearby nor incapacitate them. Maximum values are usually to be found in the relevant guidelines and regulations.

19.4.1.3 Fire Protection

There are rules regarding the fire protection of spaces, for example, in France [15], 30 min between a room and the corridor and 60 min between rooms. This is meant both for the protection of the building (to leave time for the firemen to arrive and fight the fire) and for the protection of its occupants (to leave time for orderly evacuation). On arriving on site, the firemen will check the time elapsed since the alarm went off, and should it exceed the time allocated to the protection of the structure, they will fight the blaze from outdoors.

Fire protection can be achieved through several construction techniques, for example, using concrete walls or plasterboard partitions and sealing all apertures, such as openings for cabling or piping through a wall (please note that this is a common concern with the acoustician). The sealant must, of course, be fire resistant. This makes things interesting when dealing with an expansion joint, especially a large one such as required in high seismicity areas. It is then necessary to prescribe the relevant sliding coffer with the appropriate quantity of plasterboard and mineral wool.

So far, fire protection has some nice similarities with acoustic sound insulation, for example, a love for heavy concrete walls and floors, as well as the sealing of apertures between spaces. But there may be some conflicting interests too. For example, most fire control regulations will require construction voids (e.g., a ceiling void) to be limited in size. This means that in the case of a suspended ceiling, line contacts from the required void separation walls will be added to the point contacts from the suspension, with a resulting reduced acoustic efficiency. A similar situation can be encountered on the side walls too, where the firemen will be afraid of a chimney effect. Sometimes the firemen and safety engineers may accept the idea that filling the void with mineral wool will eliminate the need for those extra separations, but beware that the wool eventually congregates at the bottom of the void and create an unwanted structural link. The worst problem is usually encountered with resiliently suspended structures. A fire occurrence can have several consequences, such as the resilient pads melting and the structure crashing down. To avoid this kind of dangerous situation, one usually has rigid heighteners (e.g., plain concrete blocks) that stand ready to receive the structure in case of failure of the pads. Similarly, there are some hooks ready to stop a resiliently suspended ceiling from crashing down in case of fire.

When dealing with a fire, one usually worries about the stability of the structure. This is a reason for the firemen in France to require a metal structure either to be visible (i.e., one can look and spot any telltale deformation due to the heat heralding the need to take cover) or to be fire protected (usually to the horror of the architect, as this often means covering a thin, elegant structure with plasterboard or simply creating a fireproof ceiling). The latter usually is to the taste of the acoustician—until the firemen require a fire detection system in the ceiling void (and this means that a trapdoor will have to be found in order to provide access for verification and maintenance of this detector) [15, 16].

Note: In older buildings (typically before the 1980s), fire protection was often performed using an asbestos coating. Whenever such a treatment is identified, one must report it to the authorities and decide [17] with the safety engineer in charge whether to remove it or hide it behind, for example, plasterwork. In the latter case, several prescriptions will probably have to be amended in order to prevent contact with the asbestos, both during the actual work and afterwards.

A word about fire doors: As opposed to common belief, a fire door is normally not an acoustic door. Sure enough, it is of sturdy construction, but its seals look nonexistent to an acoustician. Actually, the idea is for its seals to swell in case of heat elevation so as to prevent flames from getting through. If you are looking for an airproof door under normal circumstances for acoustic purposes, the fire door is not your choice.

In order to assess their fire rating performance, building materials are submitted to a fire test in a laboratory [18]. A word of advice here: The assembly of several fire rated materials will not guarantee that this assembly can be fire rated too, as the fixations (e.g., screws perforating a layer, or glue adding inflammable material) will participate in the performance. Whenever designing a complicated assembly, the advice of the safety engineer will be required.

European fire ratings are as follows [19]:

A1 (noncombustible material)
A2-s1, d0 (limited combustibility material)
B-s1, d0 (Class I surface lining)
C-s2, d0 (Class II surface lining)
D-s2, d0 (Class III surface lining)
A1fl (noncombustible floor covering material)
Cfl-s1 (Class G floor covering for exit routes)
Dfl-s1 (Class G floor covering for meeting halls and similar)
BL-s1, d0 (pipe insulation)
B1CA-S1, d0, a1 (cables)

Apart from the actual fire rating of a material, there is another issue worth considering in either high-rise buildings or underground spaces (i.e., spaces where the floor is at least 2 m under the ground level): One does not want to bring extra heating power in such spaces; therefore, assemblies using glue or wooden materials are usually not permitted. This can be a bit of a problem for the acoustician when dealing with corrective measures such as adding extra absorption on the ceiling: One cannot screw it, as this will perforate the fireproof ceiling, nor glue it due to the higher heating power involved. Whenever absorptive treatment will be needed in such spaces, the safety engineer will need to be consulted.

19.4.2 User's Behavior

Having satisfactorily designed a building is a good thing. However, even in a properly designed building, things can easily turn sour if a couple of precautions are not taken. This means that some education of the user is needed.

To start with, a fire, like all accidents, often results from a succession of minor mishaps. Here are just a few:

- Emptying the ashtray in the wastepaper basket
- Accumulating flammable material in a place without fire detection and control

- Misidentification of sensitive areas (e.g., not indicating the location of fire fighting equipment and not identifying sensitive spaces such as dangerous materials areas)
- Encumbrance of aisles and corridors
- Not checking the fire detection and fire fighting equipment on a regular basis by trained personnel
- Painting over the fire detection sensors
- Cutting off the fire warning system

Sounds ridiculous, doesn't it? Well Section 19.6.6 will give a hint of how easily things can go wrong.

19.4.3 Fire Fighting

Fire fighting is often the dream of young children. However, this is also a dangerous occupation to be left to professionals. But there are a couple of basic precautions that can be exercised by most users, provided they have been trained in the basics.

One must be aware of the location of fire fighting equipment in the building, as well as alarms.

Should a fire be suspected in a room, never attempt to tackle it alone. Instead, report the situation to the safety personnel in charge or the firemen. Then call for backup and have the fire fighting equipment ready. Try to touch the door panel. If it is hot, warn the occupants around and leave the area. If it is not, then try to touch the door handle. If it is hot, warn the occupants around and leave the area.

Do not try to enter a room whose door handle is hot: Opening the door will feed the fire and you may even be sucked inside.

If you have managed to enter, look for the source of fire. Ascertain its origin and use the relevant fire extinguishing method to put it out. There are specific classes of fire extinguishers according to the intended purpose:

- Class A is meant for standard fires (e.g., paper, cloth, etc.)
- Class B is meant for flammable liquids and gases
- Class C is meant for electrical fires (e.g., cables, etc.)
- Class D is meant for combustible metal fires (e.g., magnesium, etc.)

Do not indifferently use any class on an electric fire, as you might end up electrocuted! Only dry powder or CO_2 extinguishers (usually identified by a black band) are acceptable for such situations.

Once it seems to be extinguished, do not take for granted that the fire is out. Leave somebody to check it, and make sure it does not start again, and get combustible material out of the vicinity.

Guidance regarding basic fire fighting and hazardous materials can be found in [20].

Please do be safety minded. Remember that the temperature can rise pretty quickly, and if the fire is raging, there probably will be a flashover within 5 min of its inception, give or take a minute [21].

Remember: If you have not managed to extinguish the fire within 2 min, it probably is no longer within your capacity! Do warn the safety personnel in charge of the evolution of the situation, leave the room, and close the door (as a reminder, the occupants around should have left by now).

19.5 ANIMAL REPULSION

19.5.1 Rodent Repulsion

What has rodent repulsion to do with acoustics? The noise they generate by their squealing is usually insignificant, but it may happen (cf. Section 3.12.9). The tiny impact noise on the ceiling from their running around during the night might be a bit more uncomfortable. So what? Well, there are some areas where rodents (and eventually their predators!) will be numerous. Those lovely little animals may do such funny things as defecate on a small 1.6 V switch that may trip important mechanisms (e.g., the master switch of a power line), causing delays and problems. They may also carry potential diseases around. Next, when they are no longer heard, their predator probably is around, and it may not be a cute little kitten. As it is assumed that one does not relish the prospect of sleeping with a snake in the ceiling, there are a couple of precautions that are required:

- Make sure all apertures (e.g., for cabling and piping) are properly sealed (this will make both the acoustician and the fire engineer happy too!).
- Make sure to try to eliminate all voids during the design (this may be less fun for the acoustician, as low-frequency absorption will be more complicated to get).

In addition to such preventive measures, when confronted with a rodent invasion in sensitive spaces, there are some possibilities of rodent control through acoustic means. Good news: Those animals are sensitive to ultrasound signals (e.g., the rat emits and receives emergency signals around 33 kHz, so it will be much hampered if the background noise in these frequency bands is high). Now for the bad news: Dogs and cats will hate it, and nobody has yet found [22] what the comfort and safety limits for the human species really are. A simple trick to prevent such exposure is to connect the system to the dual-position lighting switch (when light is on there probably is occupancy, and when it is off occupants are believed out and the system can be activated). When implementing such a system, the emitters should be located so as to increase the sound level with progression inside the room (so as to convince the rodent that life is healthier outside). By the way, there is habituation with animals, so one had better use a system that emits randomly. Last, while such a system can be a help, its value will seriously be offset if there is potential food around the place, so a global effort is needed (avoid leaving readily accessible food).

19.5.2 Bird Frightening

The link between bird frightening and acoustics may seem more obvious as it usually involves audible signals. Why does one want to frighten birds? Well, for a start, they may attempt to eat the seeds painstakingly implemented in your field or garden. They may also bring various diseases (e.g., seagulls often go to the refuse heaps to feed themselves and later come to the pond). When they are strong, when wanting to nest on your roof, they may cause damage (e.g., a big gull may easily rip off a few tiles to help secure its nest!). In addition, their excreta are corrosive and will cause serious damage over the years on a metal roof.

The preferred way to frighten those birds is, when applicable, to generate the shout of a frightened bird using a special shout generator (e.g., [23]) and a loudspeaker. When a bird utters such a shout, this is a danger signal for the bird community and they flee. This works for all gulls as well as smaller birds, with the noticeable exception of the pigeon, but there are a couple of precautions that are required:

- Make sure the local regulations allow such a move (there often are limitations regarding such systems used for agricultural purposes near dwellings, but one is usually be allowed to use them with the argument that they are used twice a day at most).
- Make sure through an acoustic study that the sound level from the shout is plausible on the area to be protected, while it is minimized at the façades of the neighborhood. This will result in specifications on both the sound power level of the loudspeakers and their directivity.
- Do not use the system indiscriminately. A typical use at sunset when the birds start dozing off is usually very efficient.

Unfortunately, pigeons are not evolved enough to have a warning shout. But according to some studies, they are sensitive to ultrasound [24], so one may attempt using some of the tricks from Section 19.5.1, bearing in mind that due to the strong attenuation, the transducers generating ultrasound must be located as close as possible to the area to be protected. Other studies have reported problems with the use of ultrasound transducers producing no apparent effect on pigeons [25].

19.6 EXAMPLES

19.6.1 Vibration Control in a Seismic Environment

A cinema facility was built in the French Alps close to a railway tunnel. It was located atop of a construction featuring a commercial mall and a covered parking lot. As the commercial facility did not want to pay for vibration control measures it did not need, its upper floor was made of a 50 cm thick reinforced concrete slab that would bear the cinema facility. While the latter had its projection theatres built as a concrete structure, it was decoupled from the remainder of the building using 5 cm high resilient pads under the beams.

The problem that was quickly spotted by the safety engineer was the following: Should an earthquake occur, excitation could be much too close for comfort from the resonance of the system. Such a resonance could eventually result in the beams jumping out of the resilient supports and crashing onto the structural slab with disastrous consequences.

The problem was eventually solved using lateral pads positioned in order to quickly catch the projection theatre block in case of trouble and prevent it from moving out of its supports.

Lesson Learned: It is possible to take simultaneously both vibroacoustic and structural requirements into account as long as the subject is properly discussed between the interested parties.

19.6.2 A Thermal Bridge

In the 1970s it was fashionable to build large structures like students' residences using reinforced concrete walls and slabs. However, when the fuel crisis came, it was quickly found that too much energy was wasted through the side walls connected to the façade. A quick solution was devised using a polystyrene and plasterboard on each side of those walls. While this solved the thermal problem, it induced a significant loss in sound insulation between rooms.

Lesson Learned: Always make sure that whatever thermal cladding is proposed will not hamper the sound reduction properties of the structural elements on which they are applied.

19.6.3 Sustainable Development Project

A brand new office project was designed in France for the headquarters of a company. To keep with the spirit of its time, a sustainable development quest was carried out under the BREEAM guidelines. One of the main features called for a thermally active upper slab.

Unfortunately, BREEAM also required a minimum background noise level value in the office spaces, and the active slab would of course not be able to generate it! A few fan units were eventually required to meet the acoustic requirements.

Lesson Learned: Beware when opting for mechanical and thermal solutions that the acoustic requirements are met.

19.6.4 Outdoor Sunscreen

The glazed façades of a modern office building were treated to an outdoor cladding for sunscreening purposes.

Unfortunately, under a given incidence and wind speed, the whole assembly would start vibrating and generate a low-frequency rumble in the offices nearby. This was eventually solved through a stiffening of the structure, which also helped shift the resonance frequency of the screen assembly.

However, the sharp edges of the screen elements were still a source of whistling noise; it was strong enough to warrant some filing work to nullify this noise.

Lesson Learned: Perform an acoustic test on a sample (comprising at least one element) to make sure it does not generate too much noise.

19.6.5 Dilatation

A couple bought a partly furnished house in the south of France. On their first night, they were startled to hear violent cracking sounds coming from downstairs. Having eased their way downstairs, they did not hear anything more and concluded that the would-be intruders had fled. However, the next night the same event happened again. On calling the former owner, they were told that those loud cracking sounds actually came from the massive wooden table contracting after the daily heat period.

Lesson Learned: Objects that are sensitive to temperature variations should not be left in rooms where temperature and hygrometry may vary significantly.

19.6.6 Ashtray Emptied in the Wastepaper Basket

A worker used to empty his ashtray in the wastepaper basket at the end of his shift. True, he did take the trouble of watering the ashtray prior to disposal and there never had been a problem—that is, until one day when he came back from the holidays. After a long workday he was eagerly looking forward to going home, and having checked that no ember or smoke came from the ashtray, he emptied the contents in the wastepaper basket and left.

An hour later, his remaining colleague noticed his eyes were watering and there was a strange smell. Nothing untoward was noticed outside, but on entering the corridor to inquire, he found that there was some smoke coming from under the door. Having checked that neither door panel nor door handle was hot, he called a friend from the flat next door and then cautiously opened the door. The office was full of smoke, and there was not much left of the wastepaper basket.

Lesson Learned: Do not trust an innocuous-looking fire to be really extinguished.

19.6.7 Incorrect Disposal of Hazardous Materials

In a laboratory, some distillation had been carried out and the nasty-smelling residues now had to be disposed of. As nobody wanted the smelly stuff to be poured in the sink, the student in charge of the experiment had the bright idea to pour it in the toilets and flush it. Unfortunately, the mixture was rich in lighter-than-water flammable liquid. A few minutes later, somebody from another lab entered the toilets and threw his cigarette in the hole. One can definitely say he felt the heat!

Lessons Learned: Do not smoke in hazardous places, and do not dispose of hazardous materials in a place not identified for that purpose.

19.6.8 Insufficient Information

One night a chemistry lab was the scene of a violent fire that started with some heating plate having been left on until it eventually ignited some materials around. This properly tripped the fire detection system, and the firemen were automatically called. On arrival, the team leader noticed a small shed with the label "chemical products store" and decided at once to water it in order to prevent the fire from spreading. Unfortunately, this shed actually contained a large amount of phosphorus, and this triggered a violent fire.

Later on it was found that the building had been outfitted with fire doors, but they only went from floor to ceiling, with the ceiling void left unobstructed. This enabled the fire to spread throughout the building.

Lessons Learned: Always have a safety engineer check and prescribe the relevant prescriptions regarding fire safety. More to the point, make sure that the safety engineer participates in the commissioning of the building. Make sure all hazardous areas are properly identified. Do not leave stray inflammable material close to a source of heat, and make sure all unnecessary equipment has been switched off.

19.6.9 Negligent Behavior

One month short of commissioning, a fire erupted in the electrical room of a new hotel. It was found that the painters had stored a few cans of paint in that room, some of them left opened, together with a few cardboard boxes. A portable light that had been left switched on eventually heated the cardboard until it ignited and started a small fire that was fed by the cardboard and the paint, but it fortunately was contained by the fireproof walls of the room.

Lesson Learned: Do not store flammable materials close to a heat source.

19.6.10 Electrical Problem

Shortly before commissioning, the electricians of a large cinema facility found that there were some weird things happening on the layout. They decided to make a thorough examination of the facility after shutting off the mains. On hearing this, the HVAC contractor decided to take the opportunity to connect the air handling units (AHUs). Next thing they knew, their worker was stuck to a live electric cable. By the time they managed to pull him to safety using wooden beams and exercising the emergency procedures, that live cable had made quite a few sparks on the metallic structure of the building, until waste cardboard was ignited and the fire was fed by the bituminous waterproofing of the roof.

It turned out that in order to be able to use their equipment, some workers had made a couple of unauthorized connections that were not controlled by the master switch.

Lesson Learned: Do not make pirated electrical connections, as they are a potential source of danger for both equipment and people. Incidentally, do bear in mind that unwanted waste material may be potential fuel for fire hazards.

19.6.11 Fire Detection Tampering

The concierge of a historical building in France had experienced a few false alarms over the previous months. Fed up with the system, he managed to disconnect it. A month later there was a big demonstration in the streets and a few rockets were fired. One of them found the roof of the building, and half of the construction was gutted by the fire by the time the firemen were able to circumvent it.

Lesson Learned (at great cost): Never tamper with the fire detection and control system.

In a fortunately less dramatic example, the architect of a project decided that he did not want to see those ugly white fire detection devices on his beautiful dark ceiling and ordered the painter to paint them black. This would, of course, inhibit their fire detection capabilities. Fortunately, it was spotted by the safety engineer and proper fire detection devices were installed again.

Lesson Learned: Do not change the properties of a fire detection device, and have the system checked by a safety engineer.

19.6.12 Audibility of an Alarm

An architect tasked with the construction of a hotel was regularly reminded by his client that the budget had to be kept to a minimum. Fortunately, he managed to argue his case to have a sound alarm in each room, and the system passed the end-of-project acceptance test.

In another project, the client decided an alarm in the corridor was sufficient. When the firemen came, they remarked that due to the good sound insulation value between the corridor and room, there could be a chance of the occupants mishearing the signal, especially when listening to the radio. They consequently refused to deliver their acceptance of the project, prompting the client to install a sound alarm in each room.

In another project, the firemen decided that the signal should be heard with the shower running; this prompted an urgent tuning of the emergency sound system. A similar case happened when the firemen decided that the alarm should be heard by a patron seated on the toilet seat while water was pouring in the sink nearby, though in the latter case it eventually required the implementation of more alarm sound devices.

Lesson Learned: Always make sure the alarm is audible everywhere.

19.6.13 Obstructed Exit

Numerous fatal accidents have occurred in public receiving spaces because of obstructed emergency doors. A sadly frequent occurrence is that of dance halls where the organizers lock those doors to prevent people from entering without paying. When a fire breaks out, people often pile up against those doors until they get fatally trampled.

Lesson Learned: Never lock an emergency exit during operation.

On a funnier note, a restaurant and lounge hotel operator had managed to open a lounge with a reduced number of exits due to the presence of massive furniture. This was eventually authorized against the fire chief's recommendations on the grounds that the users would be hosted by the hotel and would be made familiar with the safety procedures of the facility. The crafty fire chief bided his time and eventually managed to get invited to a party in this lounge. Having listed the guests and made sure they were not guests of the hotel, he then wrote a nice letter to the operator complimenting him on the quality of his service, but also pointing out that as the number of emergency exits was insufficient to house parties where the guests were not clients of the hotel, it was his sad duty to require immediate administrative closure of the whole facility. The next day, the astonished clients were treated to the sight of hotel personnel frantically sawing down furniture that obstructed the other emergency exits!

Lesson Learned: Always comply with the safety requirements, and do not even think of playing cat and mouse with the firemen!

19.6.14 An Office Building without "False" Components

The architect of an office tower wanted to make a simple and efficient building. In addition, the HVAC engineer wanted to use the thermal inertia of the building. This resulted in a concept where there was no technical floor (communications between workstations were supposed to use the Wi-Fi) and no absorptive ceiling. As in France, a technical floor and a suspended ceiling are respectively known as a false floor and false ceiling; this resulted in a boast that there were no false components in the building. However, when the acoustician was eventually called on this project, he could only remark that the amount of acoustic absorption was insufficient to comply with the sustainable development standards. As no absorptive material being thermally conductive was available, one would have to use either vertical absorptive baffles or horizontal absorptive panels. But the design did not provide for the required ceiling height to allow such a scheme. Eventually, the whole construction scheme had to be reconsidered.

Lesson Learned: Do take all aspects into consideration right from the beginning of the project in order to avoid costly delays.

19.6.15 A Performance Hall

The design of the Zenith pop concert facility in Saint Etienne (by architect Foster & Partners) required significant efforts to be devoted to sustainable development. In order to comply with this goal, the HVAC engineer wanted to use the thermal inertia of the concrete walls. Of course, the acoustician also wanted a significant amount of absorption to be applied on the lateral walls to achieve a satisfactory reverberation control. This apparent contradiction eventually was solved through the use of absorptive vertical baffles secured to the lateral walls. The performance of those baffles was first estimated by calculations, and then a prototype was built and submitted to acoustic testing and architectural approval.

Lesson Learned: There usually are solutions to solve a problem as long as everybody on the design team is eager to communicate.

19.6.16 Rodent Repulsion

Rodents try to nest in sheltered and warm places. An electric cabinet can nicely suit the bill, especially when it is low voltage. Unfortunately for the operator, such cabinets usually house command circuits. In order to make sure problems are minimized, one usually introduces a minimum of three ultrasound transducers (e.g., small tweeters) in the cabinet in order to achieve at least 80 dB in the 30 to 50 kHz range. In addition, the floor void and cable tunnels are treated to the same medicine, and so is the room. In order to minimize human exposure, a common trick is to activate either the lights or the transducers using a two-way switch. Such a trick had been used by an electricity provider whose maintenance teams did not want to find themselves overly exposed to ultrasound. Inevitably, one day the light was left on (and no less inevitably, Mr. Murphy made sure it was on a Friday afternoon), so the ultrasound transducers were not activated for a couple of days; when the maintenance team came back next Monday, they were treated to the sight of their books having been eaten up!

Lesson Learned: The ultrasound rodent repulsion does work, and do not forget to switch off the light!

19.6.17 Bird Frightening

An amusement and nature park close to Paris had to close down for several weeks due to the water ponds being polluted by salmonella. The cause was found to be seagulls feeding themselves at a nearby refuse storage before coming back to sleep on the ponds. A simple system was devised using a bird scream sound generator by Merlaud [23] and a single, rather directive loudspeaker mounted on a trolley. This simple assembly enabled a worker to use it once per evening from different locations according to the wind direction and observed concentration of birds. Those birds eventually felt unsecure enough to abandon the place.

Lesson Learned: Adapt to the circumstances.

At a container handling facility, pigeon excrements were corroding the loading cranes to the extent that they had to be painted every 4 years. An ultrasound system was tentatively installed on one crane for testing purposes, and soon enough that particular piece of equipment could be spotted easily, as it was the only one left clean enough.

Lesson Learned: Ultrasound repulsion did work in this case. However, it looks like there was enough space left on other nontreated cranes that may have eased up their decision to move away.

19.7 CONCLUSION

This chapter has given an idea of how complex the situation can be, as the acoustician will have to collaborate with (at least!) the architect, the structural engineer, the HVAC engineer, and the safety engineer, not to mention the client's and end users' representatives. Close cooperation between everybody involved is essential in such matters.

REFERENCES

1. I. Ver, L. Beranek, *Noise and vibration control engineering—and applications*, J. Wiley & Sons, New York, 1992.
2. Réglementation thermique "Grenelle Environnement 2012" (French Thermal Regulation 2012, in French), Explanatory brochure, July 2010, http://www.rt-batiment.fr/fileadmin/documents/RT2012/06_07_2010_-_generalisation_des_batiments_a_basse_consommation.pdf.
3. République Française, Décret n° 2010-1269 du 26 octobre 2010 relatif aux caractéristiques thermiques et à la performance énergétique des constructions (French decree on thermal characteristics and energetic performances of buildings, in French), *Journal Officiel de la République Française*, 250 (2010).
4. Building regulations UK, http://www.kingspaninsulation.co.uk/Knowledge-Base/Building-Regulations.aspx.
5. National Insulation Association, National Insulation Training Program, 2011, http://www.insulation.org/training/nitp/.
6. http://www.ateca-fr.com/.
7. http://www.metafoam.com/.
8. Y. Lemuet, H. Peperkamp, R. Machner, Combining thermally activated cooling technology (TABS) and high acoustic demand: Acoustic and thermal results from field measurements, presented at Proceedings of InterNoise 2013, Innsbruck, 2013.
9. M. Asselineau et al., Réalisation d'un dispositif de réduction de bruit en avant d'une fenêtre (Noise reducing device in front of a window, in French), presented at Acoustics Week 87 Proceedings, Calgary, 1987.
10. J. Ph. Migneron, A. Potvin, Noise reduction of a double skin façade considering opening for natural ventilation, presented at CFA Acoustics 2012 Proceedings, Nantes, 2012.
11. E. Ph. De Ruiter, The great canyon—Reclaiming land from urban traffic noise impact zone, PhD dissertation, Delft, 2005.
12. S. Viollon, M. Asselineau, J. Ojalvo, Quelques problèmes liés à la réhabilitation (A few problems with rehabilitation, in French), presented at 8th French Congress of Acoustics Proceedings, Tours, 2006.
13. EN 60849: *Sound systems for emergency purposes*, Brussels, August 1998.
14. Cooper Notification, *Designing for intelligibility vs. audibility*, White paper, www.coopernotification.com.
15. République Française, La réglementation incendie (Fire regulation, in French), Ministère de l'Intérieur, Paris, 2013, http://www.interieur.gouv.fr/Le-ministere/Securite-civile/Documentation-technique/La-reglementation-incendie.
16. International Fire Code, 2012, http://publicecodes.cyberregs.com/icod/ifc/2012/.
17. Managing asbestos in buildings—A brief guide, HSE brochure INDG223, 2012.
18. Peutz, Fire test laboratory facilities, Commercial brochure, 2013, http://www.peutz.co.uk/index.php?sub1=facilities&sub2=fire_laboratory.
19. EN 13501-2: *Fire classification of construction products and building elements—Part 2: Classification using data from fire resistance tests, excluding ventilation services*, Brussels, 2007.
20. Community Emergency Response Team, Course, 2013, http://www.citizencorps.gov/cert/IS317/index.htm.
21. National Institute of Standards and Technology, Flashover video, http://www.youtube.com/watch?v=QqMVm72FMRk.
22. C. Howard, C. Hansen, A. Zander, A review of current ultrasound exposure limits, presented at Proceedings of Acoustics 2005, Busselton, Australia, 2005.
23. Merlaud, CSS8M bird dispersal system, Commercial brochure, Domont, France, 1995.
24. K. Lang, Bio-acoustic methods: Protection system against birds and rodents, in *Industries Alimentaires et Agricoles*, Paris, 1991, pp. 842–851.
25. M. Bomford, P. O'Brien, Sonic deterrents in animal damage control—A review of device tests and effectiveness, *Wildlife Society Bulletin*, 18, 411–422 (1990).

Glossary

Accelerometer. Vibration captor whose electric signal is proportional to the acceleration.

Acoustic absorption. Physical phenomenon turning acoustic energy into calorific energy.

Acoustic absorption coefficient. Ratio of absorbed acoustic energy to incident acoustic energy. It is noted α. Note: α theoretically ranges from 0 to 1.

Acoustic material. Material supposed to feature interesting acoustic properties (e.g., absorptive or insulating material) compared to other materials. Note: This denomination can sometimes be found with rather unscrupulous manufacturers or contractors!

AHU. Air handling unit.

AI. Articulation index. A measure of intelligibility between speaker and listener.

ALC. Articulated loss of consonants (also noted Alcons). In a good room, its value is not supposed to be higher than 15%. Note: The smaller the value, the better the intelligibility.

Ambient noise. Noise including all acoustic sources associated with a given environment, including those of interest.

Anechoic.

- Anechoic termination: Absorptive surface not generating any echo (e.g., at the end of a wave tube).
- Anechoic chamber: Room whose walls, ceiling, and floor have been absorptively treated to such an extent that there is no acoustic reflection in the frequency domain of interest.
- Semianechoic chamber: Ditto the above save for the floor, which is reflective.

Average room acoustic absorption coefficient. Equivalent absorption area of the room, divided by its total surface area. It is noted α.

Background noise. Noise generated by all acoustic sources save the one under study.

Barrier.

- Any obstacle, natural or artificial, intercepting the line of sight between the source and the receiver.
- Separating wall inside a ceiling or floor plenum.

C_{80}. Clarity. Difference, in decibels, between the sound energy received by a listener in the first 80 ms and the remaining sound energy.

Cabin. Box-like structure meant to protect the personnel from noise exposure.

CIS. Common intelligibility scale. Note: It is in the range 0 (completely unintelligible) to 1 (perfectly intelligible).

CNEL. Community Noise Equivalent Level; also known as L_{den}.

Critical distance. Maximal admissible distance between the stage and the last row of seats in a room.

Critical frequency. Frequency under which significant modes are appearing.

Critical radius. Distance from a sound source, in a room, where the direct sound pressure level and the reverberant sound pressure level have the same value.

Cutoff frequency. Lowest frequency above which the normal incidence absorption coefficient is at least 0.990.

D. Sound level difference, in decibels. Note: It is sometimes labeled bruto sound insulation.

D_c. *See* critical distance.

D_{nc}. Suspended ceiling normalized sound level difference, in decibels.

D_{ncw}. Weighted suspended ceiling normalized sound level difference, in decibels, according to ISO 717-1.

D_{new}. Sound reduction index of an air inlet.

D_{nf}. Floor or façade normalized sound level difference, in decibels.

D_{nfw}. Weighted floor or façade normalized sound level difference, in decibels, according to ISO 717-1.

D_{nTA}. A-weighted standardized sound insulation, in decibels (in some regulations since 1999).

D_{nTw}. Weighted standardized sound insulation. A single number rating of impact sound level according to ISO 717-1. Note: The higher the D or D_{nTw} value, the more insulating the construction is.

Decay rate. Rate of decrease of the indicator (pressure level, acceleration, etc.) from the instant the source generating the signal was cut.

Design team. Team in charge of designing a project. It usually is made up of such specialists as (nonexhaustive list) an architect, a structural engineer, a HVAC engineer, and an acoustician.

Diffraction. Change of direction of propagation by the edge of a surface.

Diffuse field. Environment in which the incidence of sound is equiprobable from all directions.

DL_2. Rate of spatial sound level decay, in decibels per doubling of distance (for a pink noise spectrum).

DL_2s. Rate of spatial sound level decay, in decibels per doubling of distance (for a speech noise spectrum).

Dose meter. Small portable apparatus, usually without unauthorized visualization of the result, used to measure the sound exposure level of workers.

EAA. Equivalent absorptive area.

EDT. Early decay time. Initial rate of sound decay in a room.

Emergence. Excess of particular noise over the background noise level. Note: This is also known as exceedance (United States, UK) or intrusiveness (Australia).

Enclosure. Box-like structure meant to reduce the sound energy radiated by the sound source it is shielding.

End user. Person or entity that will have the use of the project after completion.

ETC. Energy time curve. Plot of amplitude in decibels versus time.

Exceedance. Excess of particular noise over the background noise level. Note: This is also known as emergence (ISO) or intrusiveness (Australia).

Far field. Part of the sound field where the spatial decay is 6 dB per doubling of distance.

Free field. Environment free from any reflective or screening object, resulting in acoustic propagation in all directions. Note: Such an environment can be found in an anechoic room.

FSTC. Field sound transmission class, according to ASTM E413 standard.

FTL. Field transmission loss.

Fundamental. Lowest frequency of the components of a musical sound.

Hard room. Room with reflective surfaces.

Harmonics. Components of a musical sound whose frequency is a multiple of the fundamental's frequency.

Helmholtz resonator. Reactive sound absorber tuned on a given frequency, made of a volume and a channel connecting it to the aperture. Note: Perforated wooden slats on a void are an example of a Helmholtz resonator.

HVAC. Heating, ventilating, and air conditioning.

Hz. Hertz, unit of frequency. 1 Hz represents 1 cycle per second.

IEC. International Electrotechnical Commission. Note: IEC is responsible for the standards governing sound level meters.

IIC. Impact insulation class. A single number rating of impact sound level according to ASTM E492. Note: The higher the IIC value, the more insulating the test sample is.

Immission level. Sound level received at a measurement point close to the building of interest without any reflective effect (e.g., in the window opening). Note: This is used in some German and Swiss regulations.

Impulse. Very short signal (theoretically its duration is close to zero, while its level is close to the maximum).

Impulse response. Variation of the quantity of interest (e.g., sound pressure level) to an impulse. Note: In a room, the impulse response will typically be the sound pressure level versus time curve.

Insertion loss. Difference between the value measured without the noise control device and the value measured at the same receiving point with the device applied.

Intensimeter. Apparatus used to measure intensity levels.

ITD. Initial time delay gap. It represents the amount of time between the arrival of the direct sound and the arrival of the first reflection.

Intrusiveness. Excess of particular noise over the background noise level. Note: This is also known as exceedance (United States, UK) or emergence (ISO).

L_{Aeq}. A-weighted equivalent sound level.

L_{den}. Equivalent sound level over the periods day, evening, and night.

L_{dn}. Equivalent sound level over the periods day and night. Note: The start and finish times of the evening period may vary according to the local regulations in force.

L_{eq}. Equivalent sound level.

$L_{Ex,d}$. Daily noise exposure level.

L_{nAT}.

- A-weighted standardized impact sound level of a floor assembly (in some regulations until 1999). Note: The smaller the value, the better the floor assembly is.
- A-weighted standardized sound level from mechanical services (in some regulations since 1999). Note: The smaller the value, the quieter the noise is.

L_{new}. Impact sound level of a floor or floor covering (walking noise).

L'_{nTw}. Weighted standardized impact sound level of a floor under *in situ* conditions; a single number rating of impact sound level according to ISO 717-2.

L_{nw}. Weighted impact sound level of a floor under laboratory conditions, a single number rating of impact sound level according to ISO 717-2. Note: The higher the L_{nw} value, the less insulating the test sample is.

L_p. Sound pressure level.

L_w. Sound power level.

LEDE. Live end dead end. A way of acoustically treating a room with one end absorptively treated and the opposed one diffusive and reflective. Note: This is a frequent way of acoustically treating the control room of a studio.

Loudness. Subjective quantification of the magnitude of the sound.

Mass law. Law of variation of the sound reduction index as a function of the mass of the partition (doubling either the frequency or the weight will bring a 6 dB increase).

Mean free path. Average distance traveled between successive reflections in an enclosed space.

Mode.

- Resonances of a solid plate or beam.
- Resonances of an enclosed space.

Note: In rooms, the effects of modes are especially sensible at low frequencies and for small rooms.

NC. Noise criteria curve.

Near field. Zone close to the sound source, where the sound pressure level varies strongly from one position to another according to the distance to the source and its directivity.

NIC. Noise isolation class. A single number rating of the noise reduction as per standard ASTM E413.

NNIC. Normalized noise isolation class. A single number rating of the normalized noise reduction as per standard ASTM E413.

NNR. Normalized noise reduction for a reference reverberation time of 0.5 s. Note: Nowadays, one talks of standardized noise reduction for a reverberation time of 0.5 s, and normalized to an equivalent absorptive area of 10 m².

NR.

- Noise reduction, the difference between the sound levels measured or calculated at two points.
- Noise rating curve.

NRC. Noise reduction coefficient. It is the average of the absorption coefficients in the 200 to 2500 Hz third octave bands as per standard ASTM C423-09a. Note: The closer to 1, the more absorptive the material is.

Octave band. A frequency band whose upper-frequency limit is twice that of its lower-frequency limit. It is designated by its center frequency (e.g., the 1000 Hz octave band is centered on 1000 Hz and features a lower-frequency limit of 707 Hz and an upper-frequency limit of 1414 Hz). Note: There is a factor of 2 between each consecutive center frequency.

PI. Privacy index. A measure of confidentiality between rooms.

Pink noise. Noise spectrum where each frequency band has the same level (mathematically, one says the spectral density is constant). Note: This typically is the kind of spectrum used in building acoustics.

Pure tone. Sound whose spectrum features only one frequency ray without any harmonic.

R. Sound reduction index, measured in the laboratory.

R'. Apparent sound reduction index, measured *in situ*.

r_c. *See* critical radius.

R_w. A single number rating of sound reduction according to ISO 717-1. Note: The higher the R_w value, the more insulating the test sample is.

RASTI. Rapid speech transmission index. Note: This is obtained through a simplified STI measurement.

RC. Room criteria curve.

Resonance. Reinforcement associated with a natural periodicity.

Resonance frequency. Frequency where a resonance of the system occurs.

Reverberation room. Measurement room built with reflective nonparallel surfaces to reduce the occurrence of modes, featuring a high reverberation time. Note: Such a room can be used to generate a diffuse field.

RT. Reverberation time.

SAA. Sound absorption average. It is the average of the absorption coefficient in the 250, 500, 1000, and 2000 Hz octave bands as per standard ASTM C423. Note: The closer to 1, the more absorptive the material is.

Sabin. Unit of acoustic absorption. 1 Sabin is equivalent to 1 m² of a material with an absorption coefficient of 1.00. Note: Unless stated, it is implicitly assumed that this is in the MKSA Meter Kilogram Second Ampre.

SEL. Sound exposure level. The SEL represents the sound level emitted over a duration of 1 s that has the same energy as the actual noise under investigation.

SII. Speech intelligibility index. Note: It is in the range 0 (completely unintelligible) to 1 (perfectly intelligible).

Silencer. Piece of duct intended to attenuate the sound energy propagated in a ducting layout.

Single number rating. Weighted average over a defined frequency range, as per the requirements of the relevant standard (e.g., ISO 11654 for the absorption coefficient, ISO 717-1 for sound insulation, ISO 717-2 for impact sound).

SLM. Sound level meter. The handheld apparatus used to measure sound pressure levels.

SN. Signal-to-noise ratio.

SNR. Signal-to-noise ratio.

Soft room. Room with low reverberation and weak reflections.

Sound attenuation. Reduction of sound level from one point to another. Note: This can be within the same space.

Sound isolation. Capacity of a structure to prevent the intrusion of sound from one space to another. Note: Two distinct spaces are involved.

Sound level meter. *See* SLM.

SPL. Sound pressure level.

Statistical level $L_{x\%}$. Sound level value reached or exceeded during $x\%$ of the analysis time. $L_{1\%}$ can be assimilated to L_{max}, while $L_{99\%}$ can be assimilated to the background noise level.

STC. Sound transmission class. A single number rating of sound reduction or sound insulation according to ASTM E413. Note: The higher the STC value, the more insulating the test sample is.

STI. Speech transmission index. Note: It is in the range 0 (completely unintelligible) to 1 (perfectly intelligible).

STIPA. Speech transmission index for public address. Note: It is a STI adapted to the public address system.

Structure-borne noise. Noise transmitted more by means of the structure than through the separating wall, if any.

Third octave band. A frequency band that is a third of an octave band; the upper-frequency limit is approximately 1.26 times the lower-frequency limit. It is designated by its center frequency (e.g., the 1000 Hz third octave band is centered on 1000 Hz and features a lower-frequency limit of 891 Hz and an upper-frequency limit of 1112 Hz). Note: There is a factor of 1.26 between each consecutive center frequency.

TL. Transmission loss, sound reduction index.

Velocimeter. Vibration captor whose electric signal is proportional to the velocity.

Walking noise of a floor. Impact sound pressure level of a floor, measured on the emission side. Note: The higher the value, the noisier the test sample is.

Weighting. Frequency weighting is the adjustment of frequency band values according to standards requirements (e.g., IEC 851); the sum of the contributions in each frequency band gives the global weighted value. Note: Nowadays, one will frequently encounter A weighting (e.g., for environmental and occupational noise), C weighting (better taking into account the low-frequency range), G weighting (for infrasound), and Z weighting (linear weighting).

White noise.   Noise spectrum where each frequency ray has the same level. Note: This means that the sound energy will be greater in the high frequencies than in the lower ones.

α.   Absorption coefficient.

α_w.   Weighted absorption coefficient according to ISO 11654. Note: The closer to 1, the more absorptive the material is.

ΔL.   Impact sound reduction, in decibels, of a floor covering.

ΔL_w.   Weighted impact sound reduction, in decibels, of a floor covering, according to ISO 717-2.

ΔR.   Sound reduction improvement, in decibels, of a wall covering or doubling.

ΔR_w.   Weighted sound reduction improvement, in decibels, of a wall covering or doubling, according to ISO 717-1.

σ.   Radiation factor.

Index